T0245075

MatWerk

Edited by
Dr.-Ing. Frank O. R. Fischer (Deutsche Gesellschaft für Materialkunde e.V.)
Frankfurt am Main, Deutschland

Die inhaltliche Zielsetzung der Reihe ist es, das Fachgebiet „Materialwissenschaft und Werkstofftechnik" (kurz MatWerk) durch hervorragende Forschungsergebnisse bestmöglich abzubilden. Dabei versteht sich die Materialwissenschaft und Werkstofftechnik als Schlüsseldisziplin, die eine Vielzahl von Lösungen für gesellschaftlich relevante Herausforderungen bereitstellt, namentlich in den großen Zukunftsfeldern Energie, Klima- und Umweltschutz, Ressourcenschonung, Mobilität, Gesundheit, Sicherheit oder Kommunikation. Die aus der Materialwissenschaft gewonnenen Erkenntnisse ermöglichen die Herstellung technischer Werkstoffe mit neuen oder verbesserten Eigenschaften. Die Eigenschaften eines Bauteils sind von der Werkstoffauswahl, von der konstruktiven Gestaltung des Bauteils, dem Herstellungsprozess und den betrieblichen Beanspruchungen im Einsatz abhängig. Dies schließt den gesamten Lebenszyklus von Bauteilen bis zum Recycling oder zur stofflichen Weiterverwertung ein. Auch die Entwicklung völlig neuer Herstellungsverfahren zählt dazu. Ohne diese stetigen Forschungsergebnisse wäre ein kontinuierlicher Fortschritt zum Beispiel im Maschinenbau, im Automobilbau, in der Luftfahrtindustrie, in der chemischen Industrie, in Medizintechnik, in der Energietechnik, im Umweltschutz usw. nicht denkbar. Daher werden in der Reihe nur ausgewählte Dissertationen, Habilitationen und Sammelbände veröffentlicht. Ein Beirat aus namhaften Wissenschaftlern und Praktikern steht für die geprüfte Qualität der Ergebnisse. Die Reihe steht sowohl Nachwuchswissenschaftlern als auch etablierten Ingenieurwissenschaftlern offen.

It is the substantive aim of this academic series to optimally illustrate the scientific fields "material sciences and engineering" (MatWerk for short) by presenting outstanding research results. Material sciences and engineering consider themselves as key disciplines that provide a wide range of solutions for the challenges currently posed for society, particularly in such cutting-edge fields as energy, climate and environmental protection, sustainable use of resources, mobility, health, safety, or communication. The findings gained from material sciences enable the production of technical materials with new or enhanced properties. The properties of a structural component depend on the selected technical material, the constructive design of the component, the production process, and the operational load during use. This comprises the complete life cycle of structural components up to their recycling or re-use of the materials. It also includes the development of completely new production methods. It will only be possible to ensure a continuous progress, for example in engineering, automotive industry, aviation industry, chemical industry, medical engineering, energy technology, environment protection etc., by constantly gaining such research results. Therefore, only selected dissertations, habilitations, and collected works are published in this series. An advisory board consisting of renowned scientists and practitioners stands for the certified quality of the results. The series is open to early-stage researchers as well as to established engineering scientists.

Herausgeber/Editor:
Dr.-Ing. Frank O. R. Fischer (Deutsche Gesellschaft für Materialkunde e.V.)
Frankfurt am Main, Deutschland

Christian Pauly

Strong and Weak Topology Probed by Surface Science

Topological Insulator Properties of Phase Change Alloys and Heavy Metal Graphene

 Springer

Christian Pauly
Aachen, Germany

D 82 (Diss. RWTH Aachen University, [2015])

MatWerk
ISBN 978-3-658-11810-5   ISBN 978-3-658-11811-2 (eBook)
DOI 10.1007/978-3-658-11811-2

Library of Congress Control Number: 2015957085

Springer
© Springer Fachmedien Wiesbaden 2015

Printed on acid-free paper

Springer is a brand of Springer Fachmedien Wiesbaden
Springer Fachmedien Wiesbaden is part of Springer Science+Business Media
(www.springer.com)

Acknowledgments

I would like to thank all the people that contributed to this work and who gave me the necessary support in order to successfully finish my PhD thesis. First, I sincerely acknowledge *Prof. Dr. Markus Morgenstern* who gave me the opportunity to do this work within his group. His constant support as well as his permanent willingness of discussion ensured a pleasant work environment. His enthusiasm concerning all kind of scientific questions together with his broad knowledge in surface science encouraged and helped me a lot in order to improve my own scientific understanding and skills. This abilities has led to a very productive and enjoyable working environment within the whole group and which is hard to find.

Special thanks go to the *Fonds National de la Recherche (FNR) Luxembourg* which funded my research stay at the group of Prof. Morgenstern and which gave me the opportunity to start and successfully finish my PhD thesis. Especially, the very uncomplicated nature of the FNR administration staff enabled a pleasant and smooth communication process for all kind of questions.

Many thanks to my direct supervisor *Dr. Marcus Liebmann*, who helped me a lot in understanding special theoretical and experimental problems within my research field. Especially in the evaluation of my data, he was very helpful and contributed a lot to the outcome of this thesis. Further, we shared a pleasant and fruitful time in and outside the beamline during our stays at the synchrotron BESSY in Berlin.

Next, I would like to thank my former master students *Martin Grob* and *Jens Kellner* as well as my office mate *Alex Georgi* which whom I spent a pleasant time at the institute and on other occasions.
I also thank *Dr. Dinesh Subramaniam* and compatriot *Mike Pezzotta* who, especially at the beginning of my PhD time, facilitate my introduction into the field of condensed matter physics and scanning tunneling microscopy. We have stayed great colleagues and friends ever since.

For the great atmosphere in and outside the institute as well as for many helpful hands during my PhD work, I gratefully acknowledge *Dr. Marco Pratzer, Raphael Bindel, Christian Saunus, Kilian Flöhr, Nils Freitag, Silke Hattendorf, Florian Muckel, Daniel Montag, Dr. Peter Nemes-Incze, Dr. Viktor Geringer, Dr. Stefan Becker, Dr. Torge Mashoff, Sven Just, Bernhard Kaufmann, Tjorven Johnson* and *Felix Jekat*.

Special thanks go to our cooperation partners who contributed with larger and smaller assists to the successful completion of the different projects described in this thesis. The people of the group of *Prof. Dr. Oliver Rader* from the Helmholtz-Zentrum Berlin who introduced us into the field of angle-resolved photoemission spectroscopy and intensively supported us during our several stays at the synchrotron BESSY. The group of *Dr. Raffaella Calarco* and especially *Dr. Alessandro Giussani* and *Dr. Jos Boschker* from the Paul-Drude institute in Berlin for the high-quality growing and pre-characterization of the different phase change materials. *Dr. Gustav Bihlmayer* from the Peter Grünberg Institut of the Forschungszentrum Jülich for the theoretical background in the field of Sb_2Te_3 and $Ge_2Sb_2Te_5$ and his partly extremely fast help with DFT data and their interpretation. *Bertold Rasche* from the group of *Prof. Dr. Michael Ruck* in Dresden, providing us the $Bi_{14}Rh_3I_9$ samples together with a broad knowledge in the field of specific chemistry of materials and the people of the group of *Prof. Dr. Jeroen van den Brink* from the IFW in Dresden, namely *Dr. Manuel Richter* and *Dr. Klaus Koepernik* which performed the adequate theoretical calculations for this particular material.

I thank *Prof. Dr. Thomas Schäpers* for the co-evaluation of my thesis.

Further, I acknowledge *Jörg Schirra* and *Sascha Mohr* for the continuous supply of helium. Especially the Christmas helium-support requires a huge gratefulness.

Many thanks go out to the mechanical workshop under the supervision of *Peter Kordt* and *Wolfgang Retetzki* for their remarkable and quick fabrication of all kind of components with high precision and quality.

I also want to thank the administrative staff of the II. institute in person of *Margarete Betger, Beatrix Dangela, Nina Grydzhuk* and *Beate Nagel* for the easy and uncomplicated handling of all kind of administration in addition to the very pleasant and agreeable atmosphere in their office.

Last but not least, I thank *Sophie Hellinghausen* for her constant support during my PhD thesis which included good and less good times, and my parents *Josiane* and *Charles Pauly* who supported me in any way they could. Merci!

Christian Pauly

Contents

1 Introduction

In recent years, a new type of quantum matter has emerged in the field of condensed matters physics, which adds a fundamental new class of materials to the so far basic distinction between conductors and insulators. However, the classification into systems which are able to conduct an electrical current and systems which are not, is not complete. The discovery of the integer and fractional quantum Hall effect in the 1980's revealed that there exist systems which are insulating in the bulk, but necessarily exhibit a gapless edge state at the boundaries with the important characteristic that the boundary conductance is a requirement of the bulk properties of the system [1, 2]. These states have been successfully deduced by topological band theory or even more abstract topological arguments, taking into account concepts like Berry phases and Chern numbers [3, 4].

In the integer quantum Hall effect, a magnetic field is required in order to exhibit the quantized Hall conductance at the boundaries, however very recently, examples of topological phases with an insulating bulk and conducting boundary states without external magnetic field have been discovered and referred to as topological insulators (TIs) [5, 6, 7, 8, 9, 10, 11]. This new quantum state, which is also known under the terminology of quantum spin Hall state, is invariant under time-reversal symmetry and exhibits a pair of oppositely spin-polarized boundary states with spin-up and spin-down propagating in opposite directions. The helical nature, i.e. the correlation between the spin and the momentum of the edge states, is thereby due to time-reversal symmetry and the topological character of the band gap, which itself is caused by the strong spin-orbit (SO) interaction present within these systems. In detail, the SO interaction modifies the energy gap between the empty and the occupied states of insulators so that some of the states which were initially located above the gap are now lying below the gap. Analogous to the concept of the Möbius strip in mathematics, this partial inversion of the band gap can not be simply unwound without closing the gap, introducing the notion of topology into these specific insulators. From this topological point of view, a band insulator with a partially inverted band gap is topologically distinct from a band insulator with a trivial band gap which is referred to as non-trivial and trivial. The term *topological* here defines a bulk invariant (an integer) which differentiates between these distinct insulating phases and classifies them accordingly. In other words, the topological distinction between a trivial insulator and a non-trivial insulator, i.e. a topological insulator, persists in the fact that the band structure of a TI cannot be continuously transformed into that of an ordinary insulator unless a closing and a reopening of the bulk band gap takes place as long as the fundamental symmetries of the Hamiltonian describing the system remain intact. The appearance of the metallic states at the boundary can be thought of as a consequence of this transition at the interface between a non-trivial and a trivial phase. The invariant describing the different phases is a $\mathbb{Z}_2$[1] index ν and has been proposed by Kane and Mele in 2005 [8]. This new approach of topological invariants has become a powerful tool for understanding many-body phases which have bulk energy gaps and has resulted in the discovery of interesting new phases. Moreover, the new topological phases already serve as a platform for fundamental physics and might lead to technological applications like spintronics or quantum computing in the near future.

Unlike the integer quantum Hall effect, the topological insulator state can be generalized to three-dimensions (3D) with topologically protected spin-polarized surface states surrounding the surfaces of the 3D material. However, in 3D, different to the 2D case, distinct topological phases can arise. The $\mathbb{Z}_2$ topological analysis quantitatively distinguishes between materials which exhibit topological protected surface states on all of their surfaces and materials which only have surface states on some of the surfaces [12, 13]. The former is referred to as *strong* TI whereas the latter is called *weak* TI. The term *weak* here, however, is misleading as it is referring to the initial believe that the topological states in this class of TIs would be unstable with respect to most types of disorder [14, 13]. However recent theoretical work has found the opposite, namely that their surface conductivity is indeed robust with respect to disorder [15, 16, 17, 18, 19, 20, 21], and can even be stabilized by disorder, very similar to the quantum Hall effect. The identification of both topological phases in specific materials and the characterization of their respective topological properties is the major point of research within this work.

This work starts with a detailed introduction into the field of topological insulators. The integer quantum Hall effect is introduced as a first example of a topological insulator and further described in terms of topology. This approach leads us to the quantum spin Hall effect which is the basic effect of all systems studied within this work. By generalizing the quantum spin Hall effect to three dimension, the notion of strong and weak topological insulators together with their respective fundamental properties will be introduced. At the end of this introductory chapter, an overview of the major experimental research within the field is given.

---

[1]$\mathbb{Z}_2$ means that the index can only have two distinct values, e.g. 0 or 1.

In the experimental method section, the different techniques which have been employed in order to characterize the different systems will briefly be introduced. Scanning tunneling microscopy (STM) and angle-resolved photoemission spectroscopy (ARPES), which are both surface sensitive and very powerful in terms of visualizing the electronic structure of materials from the atomic to larger scale, will be described from a theoretical and experimental point of view.

The experimental part of the thesis is subdivided into two main parts. The first part deals with the identification of strong topological properties within the well-known and technologically relevant phase change materials Sb_2Te_3 and $Ge_2Sb_2Te_5$. In the field of new applications in nanoelectronics, phase change materials are of tremendous technological importance ranging from optical data storage to electronic memories. In particular, the phase change alloy $Ge_2Sb_2Te_5$ which is already widely applied for optical data storage media, such as CD's and DVD's, and might be used as non-volatile RAM in the near future too, still has a huge portfolio of undiscovered potential for further applications. The main property of these materials is the very fast and energy efficient switching between a low conductance amorphous phase and a high conductance crystalline phase on a nanosecond (ns) scale. Here, spin-sensitive photoemission spectroscopy is used in order to demonstrate the spin-momentum locking of the topological surface states of single crystal Sb_2Te_3 and show that these states form a Dirac cone with the Dirac point lying close the the Fermi level. The linear energy-momentum relation of the topological states as well as their surface character is further substantiated by the Landau level spectroscopy in scanning tunneling spectroscopy (STS). In addition, a second topologically protected surface state at lower energy exhibiting strong Rashba-type spin-splitting is identified and found to be located within a SO gap away from the Γ-point. A case which has already been predicted in 1975 and only rarely probed so far. Interestingly, the spin splitting is relatively large, e.g., larger than for typical Rashba systems like Au(111) [22, 23] or Bi(111) [24, 25, 26, 27], but lower than in Bi-based surface alloys [28]. In the case of the most prominent phase change material $Ge_2Sb_2Te_5$, which is located at the center of the pseudobinary line between Sb_2Te_3 and GeTe, theory predicts the alloy to be at the borderline of these phase change materials which exhibit topological properties. Within the experimental study, the band structure of epitaxially grown, metastable $Ge_2Sb_2Te_5$ thin films has been analyzed by ARPES and STS, such that an inverted bulk valence band close to the Fermi level could be identified confining a large band gap of 0.4 eV, which, in combination with density functional theory calculations (DFT), points to a $\mathbb{Z}_2$ topological nature of $Ge_2Sb_2Te_5$. In the special case of a phase change material, this opens up the possibility of switching between an insulating amorphous phase and a conducting topological phase on ns-time scales. Results from this part of the thesis have been published in refs. [29], [30] and [31].

The second main part of the thesis focuses on the experimental characterization of the crystal and electronic structure of $Bi_{14}Rh_3I_9$ [32], the first synthesized so-called weak TI, which is a stacked material with graphene-like layer structure but consisting of heavy atoms. Using STS at low temperature, a sub-nanometer wide edge state has been observed on the step edges of the topologically dark surface, which is the surface which exhibits no topological surface states. The edge states mapped on the top surface hereby belong to the topological surface states from the non-trivial side surfaces of the material. Thus, these edge states are a direct fingerprint of a weak TI. In addition, this particular edge state is revealed to be continuous in both energy and space within a large band gap of 200 meV, thereby, evidencing its non-trivial topology. The absence of these edge channels in the closely related, but topologically trivial insulator $Bi_{13}Pt_3I_7$ further corroborates the channels' topological nature. In contrast to other topologically protected one-dimensional states, the edge state is as narrow as 0.8 nm, making it extremely attractive to device physics. Results from this part of the thesis have been published in ref. [33].

2 Fundamentals of Topological Insulators

2.1 Introduction

In recent years, a new field in condensed matter physics has emerged, describing a novel class of bulk insulators with conducting states on their boundaries [5, 6, 7, 8, 9, 10, 11]. The driving force within these materials, which are referred to as topological insulators (TIs), is a strong spin-orbit (SO) interaction under the conservation of time-reversal symmetry (TRS). In this chapter, I will give an insight into the fundamental properties of two-dimensional (2D) and three-dimensional (3D) TIs and stress how this new class can be classified in terms of topology. This new approach of topological invariants has become a powerful tool for understanding many body phases which have bulk energy gaps, and has resulted in the discovery of interesting new phases.

In contrast to a trivial band insulator, which has an energy gap separating the valence and conduction band and thus is electrically inert, TIs have an insulating bulk but necessarily highly conducting states at the boundary. These two classes of materials are *topologically* distinct and are referred to as trivial and non-trivial. The term *topological* here defines a bulk invariant (an integer) which differentiates between these phases of matter and classifies them accordingly. In other words, the topological distinction between a trivial insulator and a non-trivial TI persists in the fact that the band structure of a TI cannot be continuously transformed by respecting the fundamental symmetries into that of an ordinary insulator unless a closing and a reopening of the bulk band gap takes place. The invariant describing the different phases is a $\mathbb{Z}_2$ index and has been proposed by Kane and Mele in 2005 [8]. The index is $\nu_0 = 1$ for 3D TIs, i.e. for bulk insulators which necessarily have robust surface states on all of their surfaces, and $\nu_0 = 0$ for all other insulators.

2.2 Insulators with metallic boundary states

A common insulator is described as a material with an energy gap separating a filled valence band and an empty conduction band. A more universal

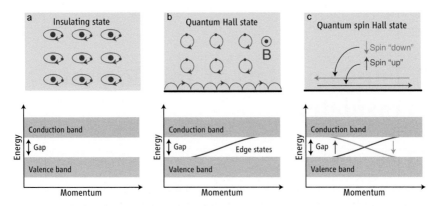

Figure 2.1: 2D insulator phases. a) Trivial insulator with electrons bound in localized orbitals. The energy gap separates the valence from the conduction band. b) In the 2D quantum Hall state, the electrons are forced into closed circular orbits by a strong perpendicular magnetic field. The quantization of the electrons leads to an insulating state inside the system. Along the sample boundary, a 1D "one way" edge state permits electrical conduction. c) The quantum spin Hall state at zero magnetic field also has a bulk energy gap but spin-polarized edge states which allow conduction at the boundary. Here, strong SO interaction plays the role of the external magnetic field in quantum Hall systems. (Adopted from [34]).

definition was given by W. Kohn describing all electronic phenomena to be local (Fig. 2.1 a)), so that the inner electrons are insensitive to perturbations from the boundaries [35]. However, the view that the existence of an energy gap guarantees the insensitivity to boundary conditions has changed with the discovery of the integer quantum Hall effect (IQHE) in a 2D electron gas (2DEG) by von Klitzing in 1980 [1]. In the quantum Hall state, the quantization of the closed circular orbits of electrons in an external magnetic field results in a bulk energy gap whereas, however, at the boundaries of the system a one-dimensional (1D) edge state appears (Fig. 2.1 b)). The unique characteristic of this edge state is that the charge flows in one direction only, making it insensitive to backscattering from impurities. Because both, band insulator and IQHE, have a bulk band gap, they appear similar from the band structure alone. The distinction between the two is a topological property which is based on the occupied bands and which is encoded in the Chern number introduced by Thouless *et al.* in 1982 [3]. In the following, I will briefly introduce the most important properties of the IQHE and its description in terms of topology. This approach is more fundamental and immediately leads to a similar phase, namely the quantum spin Hall effect (QSHE) which is the matter of study in this work. In contrast to the IQHE, the QSHE has a pair of edge states with opposite spin propagating in opposite directions (Fig. 2.1 c)). Materials which exhibit the QSHE are referred to as 2D topological insulators and can be similarly described by a topological index, distinguishing it from a trivial band insulator.

2.2.1 Integer quantum Hall effect

The quantum hall state appears when electrons which are confined to two dimensions are placed in a magnetic field which points normal to their confinement. The generated quantized orbital motion with cyclotron frequency ω_c of the electrons leads to quantized Landau levels with energy $E_n^{LL} = (n + 1/2) \cdot \hbar\omega_c$, where n is a natural number. The Landau levels are highly degenerate and if n Landau levels are filled, there is an energy gap separating the filled and empty states with the Fermi level E_F lying within the gap. The system is thus in an insulating state. Von Klitzing an coworkers experimentally realized that in such a system, which includes disorder, the longitudinal conductance σ_{xx} becomes zero while the Hall conductance σ_H, which is measured perpendicular to the position of source and drain in the experiment, exhibits quantized sequences of wide plateaus [1]. These plateaus appear at integer multiples of a fundamental nature constant, $\nu e^2 / h$, where ν is an integer and known as the filling factor. Thus, the Hall conductance is independent of geometrical details, and is even stabilized by imperfections of the materials. First, the quantization of the Hall conductance can be understood in a semi-classical picture. When the electron, which is forced on the orbital path by the strong magnetic field, comes close to the boundary, it is scattered and bounced forward along the boundary. As a consequence, it creates a conducting edge channel along the boundary. These edge channels are not affected by impurities or defects, as long as the defects are separated by more than the diameter of the orbits, as after a scattering process their sense of rotation is always the same (defined by the Lorentz force) and thus are bounced back in the former forward direction (see Fig. 2.1 b)). The edge state forms a perfect 1D conducting channel with a quantum conductance of e^2 / h. Due to the fact that there is no backscattering along the boundary, the longitudinal voltage V_{xx} becomes zero. As the discrete Landau levels form the band structure in a quantum Hall system, each filled Landau level will generate a conducting edge channel. Consequently, the number of filled Landau level, which corresponds to the filling factor ν determines the quantized Hall conductance at the edge. Thus, a quantum Hall state with quantized Hall conductance is defined, if an insulating bulk with localized electrons is present, such that only the edges which have a series of perfectly propagating edge channels conduct currents.

2.2.2 The Hall conductance as a topological invariant

The pioneering work of Thouless *et al.* [3] (TKKN) and Laughlin *et al.* [36] introduced a fundamental new understanding of this quantum state of matter, as the quantum Hall state provided the first example of a system which is topologically distinct from all other states of matters known before. The quantum Hall effect thus defines a topological phase in the sense that certain fundamental characteristics, like the quantized Hall conductance or the number of perfectly propagating edge channels, are insensitive to smooth changes in the parameter space. In mathematics, topological classification focuses on the fundamental distinction of shapes. From this point of view, a coffee cup and a donut are topologically equivalent, as both have exactly one single hole and may be converted inside each other by a "smooth" deformation, i.e., a deformation process which does not need the violent action of creating a hole. In physics, the analogy consists in the concept that Hamiltonians $H(k)$ describing states with gapped band structures are topologically equivalent if $H(k)$ can be continuously deformed into one another without closing the gap. These classes are distinguished by a topological invariant called the Chern number n ($n \in \mathbb{Z}$).

Quantized Hall conductance by Thouless, Kohmoto, Nightingale and den Nijs

Thouless *et al.* for the first time defined an invariant in their original work, which describes the quantization of the Hall conductance in the non-interacting IQHE in integer multiples, called the TKKN integer [3][1]. They considered a system of non-interacting electrons moving in an xy plane on a lattice perpendicular to a uniform magnetic field **B**. The system is described by the following Hamiltonian:

$$H = \frac{1}{2m}(p + eA)^2 + U(r), \tag{2.1}$$

where A is the magnetic vector potential and $U(r)$ representing a periodic potential of the lattice which satisfies U(x+a_1,y) = U(x,y+a_2) = U(r), with a the lattice constant. However, the Hamiltonian is not invariant under a translation along x or y because the vector potential A is not periodic within the lattice. A translation operator $T(R) = \exp((i/\hbar)R \cdot p)$, which shifts an arbitrary function by a lattice vector $R = na_1 + ma_2$, transforms the magnetic vector potential into $A(r+R)$, which is different to $A(r)$. Thus, $T(R)$ does not commute with H.

---

[1]The following derivation of the quantized Hall conductance as integer multiple is lean on the work of Kohmoto [37] and Watson [38], which are both based on the original work of Thouless *et al.* [3]

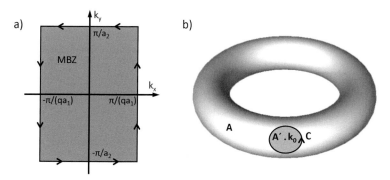

Figure 2.2: a) Magnetic Brillouin zone (MBZ) constructed by the magnetic unit cell (not shown here) through which a total of q quanta of magnetic flux passes. MBZ is restricted by $k \in [0, 2\pi/(qa_1)] \times [0, 2\pi/a_2]$. The arrows mark the direction of the line integral in eq. 2.6 which moves around the boundary of the zone and which is proportional to the conductance of one subband. b) Transformation of the MBZ into a torus. A vector potential $A'(k)$ (shaded area) is defined by a singularity of the function $u_k(r)$ at k_0. The phase mismatch line C between the vector potentials $A'(k)$ and $A(k)$ is marked. (Adopted from [38]).

However, an appropriate gauge transformation can make the Hamiltonian invariant. Consider magnetic translation operators of the form

$$T_B(R) = exp((i/\hbar)R \cdot [p + e(r \times B)/2])), \qquad (2.2)$$

which commutes with the Hamiltonian but do not commute with each other since [37]:

$$T_B(a_1)T_B(a_2) = exp(2\pi i\alpha)T_B(a_2)T_B(a_1). \qquad (2.3)$$

Thereby $\alpha = (eB/h) \cdot a_1 a_2$ and equals the number of magnetic flux quanta Φ_0 passing through each unit cell of the lattice. Thus, when α is a rational number, the magnetic translation operator commute with each other. Consequently, one can define a new unit cell which contains an integer number of magnetic flux quanta and is built by an integer number of unit cells without magnetic field. This so-called magnetic unit cell can be formed by the vectors qa_1 and a_2 with the primitive lattice vector taking the form $R = n(qa_1) + ma_2$ so that q magnetic flux quanta are in the unit cell. In this gauge, the eigenstates $\psi_{\lambda k}(x,y)$ of the above Hamiltonian can be labeled by a discrete band index λ and a momentum k in the magnetic Brillouin zone, which is restricted by $k \in [0, 2\pi/(qa)] \times [0, 2\pi/a]$ (Fig 2.2 a)). These are the Bloch wave-functions

$$\psi_{\lambda k}(r) = e^{ik \cdot r} u_{\lambda k}(r), \qquad (2.4)$$

consisting of a plane wave $e^{ik\cdot r}$ and a function $u_{\lambda k}(r)$ which is periodic in the magnetic unit cell.

For a system of non-interacting particles like the one described by the Hamiltonian in 2.1, Thouless et al. [3] derived a formula for the Hall conductance σ_H which is based on the so-called Kubo formula. The expression is deduced from a second-order perturbation theory as a linear response of the system to a small applied electric field [39]. It follows from the Kubo formula, that the transversal velocity of the particles in an electric field within the system is a property of the periodic part of the Bloch wave-function (see ref. [37] for a full derivation) so that the Hall conductance σ_H can be determined as a property of the periodic function $u_{\lambda k}(r)$:

$$\sigma_H = \frac{e^2}{h} \sum \int d^2r \int \frac{d^2k}{2\pi i} \left(\frac{\partial u_{\lambda k}(r)^*}{\partial k_x} \frac{\partial u_{\lambda k}(r)}{\partial k_y} - \frac{\partial u_{\lambda k}(r)^*}{\partial k_y} \frac{\partial u_{\lambda k}(r)}{\partial k_x} \right), \quad (2.5)$$

where the sum is over the occupied subbands indexed by λ, and the r and k integrals over the magnetic unit cell and magnetic Brillouin zone, respectively. Before I discuss the topological aspect of this equation, I first briefly revise Thouless' direct proof that the Hall conductance implies integer quantization. For the sake of simplicity, one looks at the contribution of one single occupied band to the Hall conductance which is defined by σ_λ. Together with the observation that the integrand in 2.5 can be written as the z-component of the curl of some vector field, it then follows:

$$\begin{aligned}
\sigma_\lambda &= \frac{e^2}{h} \int d^2r \int \frac{d^2k}{2\pi i} \left(\frac{\partial u_k(r)^*}{\partial k_x} \frac{\partial u_k(r)}{\partial k_y} - \frac{\partial u_k(r)^*}{\partial k_y} \frac{\partial u_k(r)}{\partial k_x} \right) \\
&= \frac{e^2}{h} \int d^2r \int \frac{d^2k}{2\pi i} [\nabla_k \times (u_k(r)^* \nabla_k u_k(r))]_z \\
&= \frac{e^2}{h} \int d^2r \oint \frac{dk}{2\pi i} (u_k(r)^* \nabla_k u_k(r)).
\end{aligned} \quad (2.6)$$

The third line uses Stokes' theorem to rewrite the integral as a line integral around the magnetic Brillouin zone. The integration contour is depicted in Fig 2.2 a). Thus, one ends up with an expression for the Hall conductance which depends on the change of the periodic function $u_k(r)$ along a closed loop around the magnetic Brillouin zone. As corresponding points on the boundary of the magnetic Brillouin zone represent the same physical state (for example k_y values differing by $2\pi/a_2$), the respective function $u_k(r)$ at opposite points only differs by a total phase factor which is independent of r:

$$\begin{aligned}
u_{k_x, \pi/a_2}(r) &= e^{i\Theta(k)} u_{k_x, -\pi/a_2}(r), \\
u_{\pi/(qa_1), k_y}(r) &= e^{i\Theta(k)} u_{-\pi/(qa_1), k_y}(r),
\end{aligned} \quad (2.7)$$

for the horizontal and vertical lines of the magnetic Brillouin zone, respectively (Fig 2.2 a)). Further, as $u_k(r)$ is periodic within the magnetic unit cell, the function can be normalized, with the r integral of $u_k(r)$ over the magnetic unit cell set equal to unity:

$$\int d^2 r \, |u_{\lambda k}(r)|^2 = 1. \tag{2.8}$$

It then follows for the Hall conductance σ_λ:

$$\begin{aligned}
\sigma_\lambda &= \frac{e^2}{h} \oint \frac{dk}{2\pi i} (i\nabla_k \Theta(k)) \\
&= \frac{e^2}{h} \oint \frac{dk}{2\pi} \nabla_k \Theta(k).
\end{aligned} \tag{2.9}$$

Thus, the total line integral corresponds to the total change of the phase $\Theta(k)$ of the function $u_k(r)$ around the boundary of the magnetic Brillouin zone. The change of the phase around a closed loop must be an integer multiple of 2π, so that the Hall conductance for each completely occupied band must consequently be an integer:

$$\sigma_\lambda = \frac{e^2}{h} \cdot \nu. \tag{2.10}$$

The integer ν is known as the TKKN integer and reflects the quantized value of the Hall conductance.

In the following, the Hall conductance will be analyzed in terms of topology and the integer found by Thouless *et al.* is shown to be equivalent to the topological invariant of the class of $U(1)$ fiber bundle, namely the Chern number.

Equivalence of the TKKN invariant and the Chern number

Mathematically, topological invariants like the Chern number are associated to distinct Chern classes which characterize specific fiber bundles, as for example the $U(1)$ fiber bundle. Topological invariant here means that equivalent bundles are described by the same number. It was first recognized by Avron *et al.* [40] that in the case of non-interacting electrons the periodic part of the magnetic Bloch wave-functions $u_{\lambda k}(r)$ in the TKKN description form such a $U(1)$ fiber bundle on the base manifold of a torus T^2. An essential observation thereby is that the magnetic Brillouin zone is topologically a torus T^2, as two k_x-values which differ by $2\pi/(qa_1)$ are equivalent, as are also two k_y-values which differ by $2\pi/a_2$ (cf. Fig. 2.2). The integral in eq. 2.6 is then equivalent to a surface integral over some curvature of the torus T^2 and defined as a topological invariant, namely the Chern number of the $U(1)$ bundle

[40, 41]. In order to substantiate the above considerations, let us go back to the expression for the Hall conductance for a single band (2nd line of eq. 2.6):

$$\sigma_\lambda = \frac{e^2}{h} \int \frac{d^2k}{2\pi i} \nabla_k \times \left(\int d^2r \cdot u_k(r)^* \nabla_k u_k(r) \right)_z , \qquad (2.11)$$

and define a fictitious vector potential in momentum space which is based on the functions $u_k(r)$ and has the form:

$$A(k) = \int d^2r \cdot u_k(r)^* \nabla_k u_k(r) = \langle u_k | \nabla_k | u_k \rangle . \qquad (2.12)$$

$A(k)$ is also known as the Berry vector potential and $F(k) = \nabla_k \times A(k)$ the corresponding Berry curvature. Thus, the Hall conduction can be written as the surface integral of the Berry curvature:

$$\sigma_\lambda = \frac{e^2}{h} \cdot \frac{1}{2\pi i} \int d^2k \, (\nabla_k \times A(k))_z . \qquad (2.13)$$

Applying Stokes' theorem, the above equation is equivalent to the line integral over the Berry vector potential along the boundary of the magnetic Brillouin zone, which however is topologically a torus and thus has no boundary. Consequently, $\sigma_\lambda = 0$ if $A(k)$ is uniquely defined on the entire torus T^2.

As the vector potential is completely defined by $u_k(r)$, $A(k)$ is smooth over T^2 if there are no values of k where $u_k(r)$ vanishes. Hence a non-zero Hall conductance appears when there is a zero somewhere in the functions k-dependence. As the magnetic flux is always accompanied by a zero in the wave function, there must necessarily be an integer number of zeros in the magnetic Brillouin zone. So no matter how one tries to define a vector potential, one must always end up with a singularity, which is the topological equivalence of a non-trivial $U(1)$ fiber bundle.

Hence, one looks at a single zero of $u_k(r)$ at some random point k_0 on the torus, and define a new vector potential $A'(k)$ around k_0. The situation is illustrated in Fig. 2.2 b). As the functions $u_k(r)$ are defined only up to a phase factor, there must necessarily be a phase mismatch line (marked by C in Fig. 2.2 b)) by the clash of phases coming from the zero (defined by $A'(k)$) and from the periodic boundaries (defined by $A(k)$). As a consequence, the functions of the two areas differ at the mismatch line by an r-independent phase factor:

$$u_{k'}(r) = e^{i\Phi(k)} u_k(r), \qquad (2.14)$$

so that the corresponding vector potentials differ by:

$$A'(k) = i\nabla_k \Phi(k) + A(k). \qquad (2.15)$$

The Hall conductance may now be written as the sum of the integrals of both curvatures over their respective areas of the torus according to eq. 2.13. Additionally applying Stokes' theorem, it follows:

$$
\begin{aligned}
\sigma_\lambda &= \frac{e^2}{h} \cdot \frac{1}{2\pi i} \int_U d^2k \left(\nabla_k \times A(k)\right) + \frac{e^2}{h} \cdot \frac{1}{2\pi i} \int_{U'} d^2k \left(\nabla_k \times A'(k)\right) \\
&= \frac{e^2}{h} \cdot \frac{1}{2\pi i} \oint_C dk A'(k) - A(k).
\end{aligned}
\tag{2.16}
$$

One can immediately see that by inserting eq. 2.15, the Hall conductance σ_λ only depends on the change of the phase $\Phi(k)$ around the mismatch line C. However, as the phase is a single-valued function of k, it follows that its change around a closed loop is an integer multiple of 2π. Consequently, one gets:

$$
\sigma_\lambda = \frac{e^2}{h} \cdot n,
\tag{2.17}
$$

with n being the integer Chern number. This shows that the IQHE can be assigned to a non-trivial class of state and that the TKKN integer is indeed a topological invariant. Moreover the value of n depends on the topology of the mismatch line and only if the mismatch line is not contractible on the torus, n is different from zero and the system thus non-trivial. One major consequence of this result implies that the quantization of the Hall conductance is insensitive to experimental details, such as material used or shape of the sample, as only geometrical considerations of the band structure of Landau levels in a periodic potential have led to the above result. It can even be shown, that the result remains robust, if weak disorder or electron-electron interaction is included in the Hamiltonian.

Bulk-edge correspondence

An important consequence of the IQHE are the dissipationless edge states which move in opposite directions for opposite boundaries. The non-trivial nature of these states as well as their existence are deeply connected to the topology of the bulk quantum Hall state. Imagine the interface between a quantum Hall state ($n = 1$) and a trivial insulator, such as vacuum, which Chern number is necessarily $n = 0$ since there are no occupied states. At some point along the interface, the energy gap has to vanish in order for the Chern number to change from trivial to non-trivial or vice-versa. Thus, the system must exhibit exactly n transfers of extended states through E_F bound to the region where the energy gap passes through zero, as long as the symmetry that protects the bulk insulating state is not broken by the boundary.

A more rigorous argument, however, for the existence of edge states in the quantum Hall state responsible for the Hall current is given by Laughlin [36]. In his original work, Laughlin describes a cylindrical quantum Hall sample with an insulating bulk through which an integer multiple of flux quanta passes. This flux drives a charge pump from one side of the sample to the other without the introduction of energy to the system. Laughlin thereby introduced the concept of charge polarization, which defines whether charge is effectively transported through the system or not. In the case of the IQHE, Laughlin found a charge transfer from one boundary to the other through the insulating bulk sample, so that there must exist edge states at each boundary at E_F ready to receive or donate electrons. Moreover, the number of edge states driving the Hall current at both boundaries must be different so that a current is effectively flowing in the system. Thus, the difference between left boundary states N_l and right boundary states N_r is determined by the topological structure of the bulk states which define the Chern number [9]:

$$N_l - N_r = \Delta n, \qquad (2.18)$$

with Δn the difference of the Chern number across the boundary. This relation summarizes the bulk-edge correspondence. Note that in the topological description of the QSHE (chapter 2.2.4), an analogous argument to the one of Laughlin is used in order to explain the existence of the non-trivial spin-polarized edge/surface states.

2.2.3 Quantum spin Hall effect or time-reversal invariant 2D topological insulator

The disadvantage of quantum Hall systems are their demanding requirements, such as a strong magnetic field and very low temperature. The quest in recent years was to find a class of materials having the same robust edge states without the need of a magnetic field and stable up to room temperature. Kane and Mele [8, 42] proposed in 2005 a spin version of the quantum Hall effect, namely the quantum spin Hall effect (QSHE), with an intrinsic SO interaction as the driving force, so that a non-trivial bulk energy gap is induced in the material resulting in a pair of gapless spin filtered states at the boundary. Whereas in the case of the IQHE, the time-reversal symmetry (TRS) is explicitly broken by the presence of a magnetic field, this new quantum state belongs to a class which is invariant under TRS. In contrast to the IQHE, the edge states in the QSHE are spin-polarized, so that electrons with spin-up propagate in one direction an electrons with spin-down move in the opposite direction (Fig. 2.3 a)). The helical nature, i.e. the correlation between the spin and the momentum of the edge states is thereby topologically required by the

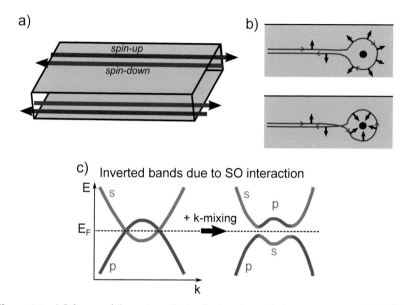

Figure 2.3: a) Scheme of the spin-polarized edge channels in a quantum spin Hall insulator. Every edge hosts a pair of spin-polarized states with spin-up and spin-down channel propagating in opposite directions. b) Scheme illustrating the absence of backscattering of the edge states at a time-reversal invariant impurity. The electron can take two possible paths around the impurity when scattered, either rotating its spin clockwise or anticlockwise about 180°. A geometrical phase factor of the spin between the two paths leads to a destructive interference, thus prohibiting backscattering (adopted from [10]). c) Band structure of a model quantum spin Hall insulator visualizing the effect of SO coupling. SO coupling pushes the p-type valence band above the s-type conduction band so that an inversion of bands takes place at the Fermi level E_F for the center points in k-space. An interaction term leads to a symmetric and antisymmetric combination of the two bands at the crossing points. Thus, a band gap occurs together with a changing of s/p character (as marked) of the two bands.

spin Hall conductance, with the exact relative directions depending on the SO interaction within the system. TRS requires that the quantized charge transported by each edge channel is equal (Hall conductance for one spin channel $\sigma_{xy} = e^2/h$ and for the other $\sigma_{xy} = -e^2/h$) so that the overall charge Hall conductance is zero. However, spin charge is transported from one side to the other so that the quantized spin Hall conductivity is nonzero. This means, that even though the TKKN invariant ν for this system is zero (no Hall conductance), the ground state of the QSHE is different from that of a trivial insulator. This implies that there must be a similar topological classification for time-reversal invariant systems distinguishing systems with trivial spin Hall conductivity and such with non-trivial one. This invariant is called a $\mathbb{Z}_2$ topological index ν [8] and can be calculated for the occupied Bloch states of a system with a bulk band gap, analogous to the TKKN invariant, in order to distinguish between a quantum spin Hall phase ($\nu = 1$) and a trivial insulator ($\nu = 0$).

The outstanding property of a quantum spin Hall insulator, i.e. a 2D topological insulator (TI) is its perfect conductance transported by the edge states. In contrast to ordinary conductors, which have both spin-up and spin-down states moving in both directions and are thus susceptible to Anderson localization [43, 44], the helical nature of the 2D TIs edge states prevents them from localization [45]. A semiclassical argument for the lack of backscattering in the case of time-reversal invariant impurities was given by Qi and Zhang [10] as visualized in Fig. 2.3 b). A forward moving spin-up state at the edge of the 2D TI can either make a clockwise or anticlockwise turn of its spin around an impurity. As the electrons which propagate backward have a spin-down state, the electron spin has to rotate adiabatically either by an angle of π or -π when rotating around the impurity. As the two paths therefor differ by a 2π rotation of the electron spin, changing the spin phase by π, the two paths in Fig. 2.3 b) interfere destructively, thus leading to perfect transmission. Hence, only a magnetic impurity which is able to break time-reversal symmetry, may lead to localization. Thus, the spin-polarized edge states in the quantum spin Hall insulator are protected from backscattering by TRS.

Beside TRS, it is obvious that in a system with spin-degenerate states where the momentum, i.e. the movement of electrons, is opposite for two different spin components, SO interaction plays an important role. SO coupling leads to a splitting of spin states in energy and is strongest for bands where the periodic part of the Bloch wave-function $u_{\lambda k}(\mathbf{r})$ exhibits an orbital quantum number, as for example in p-type or d-type bands, and does not exist for bands without, e.g. pure s-type bands. As most of the semiconductors have a p-type valence band and a s-type conduction band, it follows that the spin components of the valence band shifts much more strongly than the conduction band. Thus, if the SO coupling is strong enough, a crossing of the p-type valence band and the s-type conduction band may occur as displayed in Fig. 2.3 c). At the k-points where the bands cross each other, perturbation theory demands the formation of a symmetric and antisymmetric combinations of p- and s-type states with different energy, typically accompanied by a gap opening at these points. Thus, the inversion of bands due to SO interaction leads to a band gap where the band character of the inner part between the anticrossing points is inverted with respect to the outer part. This has important consequences for the total inversion symmetry of the occupied states at specific k-points. Since the s-bands are even and the p-bands odd under spatial inversion, the overall parity for the occupied bands in the inner part differs from the parity in the outer part. Later, we will see that this change of parity at only specific k-points in the Brillouin zone induces a phase transition in which the $\mathbb{Z}_2$ invariant changes from trivial to non-trivial. This is the base for the appearance of the topological protected boundary states.

The first experimental quantum spin Hall state in HgTe/CdTe quantum well structures

Kane and Mele [42] proposed the QSHE for graphene [46, 47] but the weakness of the SO coupling in the material made the suggestion fail, as the gap opened by SO coupling turned out to be extremely small (of the order of 10^{-2} meV) [48]. As the strength of the SO coupling within a material increases with the charge number of the atoms constituting the system, an approach for the search of the first 2D TI was to look for semiconductor materials built by heavy elements. The experimental hallmark for the discovery of the QSHE is the quantization of the longitudinal conductivity σ_{xx} in integer multiples of $2e^2/h$. This results from the fact, that in contrast to the IQHE, if a current is applied from left to the right in Fig. 2.3 a) both edges contribute to the current, namely the spin-up channel on the upper edge and the spin-down channel on the lower edge.

In 2006, Bernevig *et al.* [5] proposed quantum well structures of $Hg_{1-x}Cd_x Te$ to exhibit the QSHE. The structure consists of a thin layer of HgTe sandwiched between crystals of CdTe (Fig. 2.4 a)). HgTe and CdTe and their alloys have strong SO interactions and are a well-studied family of semiconductor materials. CdTe has a trivial band structure similar to other semiconductors with a relatively large band gap (1.56 eV) formed by a p-type valence band and a s-type conduction band. In HgTe, however, the strong SO coupling which is induced by the heavy element Hg, leads to an inverted band gap of 0.2 eV with the p-level lying above the s-level (right scheme in Fig. 2.4 a)). Thus, HgTe was predicted to be a 2D TI and to exhibit a quantized conductance of $2e^2/h$. In the experiment proposed by Bernevig *et al.* [5] the thickness of the HgTe layer is chosen very small ($d < d_c = 6.3$ nm) so that confinement effects play a role. The additional confinement energy to the bands is positive for the s-type band and negative for the p-type band, such that the bands reinvert to the normal regime, with the p-type band forming the valence band and the s-type band forming the conduction band. The normal regime is displayed in the left image of Fig. 2.4 a). If the thickness of the HgTe layer exceeds d_c, the system transforms again in the inverted topological regime (right scheme in Fig. 2.4 a)).

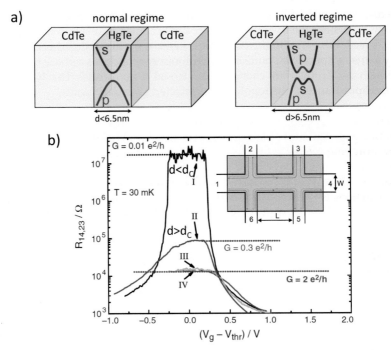

Figure 2.4: a) HgTe/CdTe quantum well structure with different thicknesses of the HgTe layer. The corresponding valence and conduction bands are marked. For $d < 6.5$ nm, HgTe is in the normal regime with a p-type valence band and a s-type conduction band. For $d > 6.5$ nm, the bands cross and HgTe is in the inverted regime which characterizes the quantum spin Hall state. b) Experimental longitudinal four-terminal resistance $R_{14,23}$ as a function of the gate voltage for the normal ($d = 5.5$ nm) (I) and the inverted ($d = 7.3$ nm) (II, III and IV) quantum well structures measured at zero magnetic field and $T = 30$ mK. The gate voltage tunes E_F through the bulk gap. The device sizes are $(20.0 \times 13.3)\,\mu m^2$ for sample I and II, $(1.0 \times 1.0)\,\mu m^2$ for sample III, and $(1.0 \times 0.5)\,\mu m^2$ for sample IV. Sample I ($d < d_c = 6.3$ nm) shows an insulating behavior, while sample III and IV reveal a nearly perfect quantized conductance of $2e^2/h$ of the QSHE. Inset shows a scheme of the sample layout in the original experiment of König *et al.* [7] with the ohmic contacts labeled 1 to 6, respectively. Red and blue arrows mark the counterpropagating spin-polarized edge channels of the quantum spin Hall state. (Adopted from [7]).

It was only soon after this proposal that König *et al.* [7] provide the experimental proof of the first quantum spin Hall insulator by measuring the low-temperature ballistic edge transport for the two devices. Figure 2.4 b) shows the resistance measurements for different samples as a function of gate voltage tuning E_F through the bulk energy gap. Sample I is a narrow HgTe quantum well ($d = 5.5$ nm) that is in the normal regime, thus having no topological edge states in the bulk band gap. And indeed, the measurement revealed a very large resistance in the voltage range of the band gap, increasing up to 20 MΩ, which is the maximum resistance that could be detected within this particular experiment. Sample II, III and IV are wider wells ($d > 6.3$ nm) and thus are in the inverted regime. Sample III and IV indeed show the nearby exact conductance of $2e^2/h$ confirming the 2D topological nature of HgTe. Both samples have the same length ($L = 1\,\mu$m) but different widths ($w = 0.5$ and $1\,\mu$m), proving that the transport occurs at the edge (the geometrical details of the measurement are displayed in the inset of Fig. 2.4 b)). Note that the deviation of the perfect conductance value of $2e^2/h$ for sample II ($L = 20\,\mu$m) is due to finite temperature scattering effects. The former experiment successfully demonstrate the existence of the quantized conductance of the boundary states which is the fingerprint of the QSHE. The experimental results further provide indirect evidence for the thesis that changing the topology indeed requires that the gap between the inverted and normal region (Fig. 2.4 a)) of the band structure has to be closed. In the following chapter, the concept of the time-reversal polarization, which characterizes time-reversal invariant Hamiltonians and from which one can derive the $\mathbb{Z}_2$ topological invariant will be introduced. Later, its generalization to three dimensions will be discussed.

2.2.4 Construction of $\mathbb{Z}_2$ invariant for 2D topological insulators

In the course of the first prediction of the QSHE in graphene [42], a large number of mathematical formulations of the topological $\mathbb{Z}_2$ invariant have been developed [8, 12, 49, 14, 50, 51, 52, 53, 54, 55]. One of these concepts was given by Fu and Kane [12] and provides a rather physical meaning for the mathematical construct of the $\mathbb{Z}_2$ topological index. It introduces the idea of time-reversal polarization (TRP), which defines whether a net spin is transported from one edge to the other or not, without adding energy to the system. This requires that in an insulating bulk system, if a spin transfer takes place and the bulk Hamiltonian is only changed adiabatically, i.e. ending with the same Hamiltonian as at the beginning, there must necessarily be edge states at the Fermi level E_F. The time-reversal polarization by itself is not meaningful, however changes in the time-reversal polarization due to adiabatic changes in the bulk Hamiltonian are well defined and determine the $\mathbb{Z}_2$ invariant.

This way of looking at the $\mathbb{Z}_2$ invariant and its direct implication of a nontrivial or trivial system is in close analogy to the approach of the charge po-

a)   b)

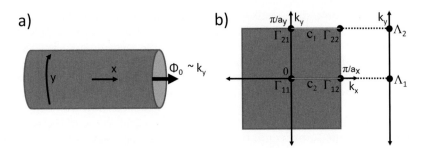

Figure 2.5: Fu and Kane concept of a spin pump in the quantum spin Hall state. a) A 2D cylinder with a circumference of a single lattice constant is threaded by a magnetic flux Φ in x-direction. Φ plays the role of the edge crystal momentum k_y in a 2D band structure. b) The magnetic fluxes $\Phi = 0$ and $\Phi = h/2e$ correspond to edge time-reversal invariant momenta Λ_1 and Λ_2 with Λ_i being the projection of pairs of the four bulk time-reversal momenta Γ_{ij} originating from the 2D Brillouin zone as marked. The lines c_1 and c_2 are used to calculate the time-reversal polarization in the real space x-direction P_x^{TR} for the two time-reversal invariant point Λ_1 and Λ_2. (Adopted from [14]).

larization which was used by Laughlin [36] to topologically define the quantization of the Hall conductance in the IQHE. The change in the charge polarization thereby characterizes the charge differences that is pumped from one edge to the other describing whether a system is in the quantum Hall phase or not. It thus provides a physical meaning to the Chern number. In Laughlin's gedanken experiment, a quantum of magnetic flux is adiabatically inserted into a cylindrical quantum Hall sample, whereas the change in the magnetic flux throughout the cylinder drives the quantum pump, as an increase of the flux results in a transfer of charge from one side to the other in the cylindrical sample. In general, the eigenstates of the system before and after the variation of exactly one flux quantum $\Phi_0 = h/e$ are identical so that the Hamiltonian is gauge invariant under flux changes of integer multiples of $\Phi_0 = h/e$. It thus describes a cycle of pump, i.e. a charge transfer without the introduction of energy to the system. Laughlin derived that the total charge transported through the cylinder during one pump can be interpreted as the change in the charge polarization which is exactly an integer multiple of e. Thus, the change in charge polarization is quantized and precisely characterized by the Chern number.

Fu and Kane [12] adopted the idea of charge polarization and applied it to a system with time-reversal symmetry, in order to define the QSHE. They considered a system which consists of a two-dimensional cylinder with a circumference of a single unit cell in y-direction and a finite length in x-direction

(Fig. 2.5 a)). The magnetic flux threading the cylinder along the x-axis gener-
ates the spin pump and plays the role of the wave vector k_y in the 2D Brillouin
zone (Fig. 2.5 b)), as changing the flux by exactly one flux quantum $\Phi_0 = h/e$
introduces a phase shift of 2π per unit cell in y-direction and thus increasing
the wave vector by $2\pi/a_y$. The increase of magnetic flux from $0 \rightarrow$ to Φ_0 thus
corresponds to a full adiabatic cycle evolution and defines the change in the
so-called time-reversal polarization. If there is no change in the polarization,
no spin has effectively been pumped and the system is trivial. In the case of
a time-reversal invariant system, a flux quantum of $\Phi = 0$ and $\Phi_0/2 = h/2e$
belongs to the two edge time-reversal invariant momenta $k_y = \Lambda_1$ and $k_y =$
Λ_2 which are the projections of pairs of the four bulk time-reversal invariant
momenta Γ_{ij} of the 2D Brillouin zone as displayed in Fig. 2.5 b). These two
points define half of a spin pump cycle and decide whether there is a change
in the time-reversal polarization, i.e. a spin transported from one edge to the
other in the x-direction of the cylinder, when k_y is moved from the center of
the Brillouin zone (Λ_1) to the boundary (Λ_2).

Time reversal symmetry and the $\mathbb{Z}_2$ invariant

In the following, the situation described above will be analyzed more for-
mally[2]. Fu and Kane proposed a time-reversal symmetric spin $1/2$ electronic
model (detailed description in ref. [12]) in order to describe a 2D TI. Time-
reversal symmetry is represented by an anti-unitary operator and can be writ-
ten as $\Theta = \exp(\frac{i\pi}{\hbar} S_y) K$, where S_y is the spin operator and K the complex
conjugation. The system is then described by:

$$H(-\mathbf{k}) = \Theta H(\mathbf{k})\Theta^{-1},\qquad(2.19)$$

Since $\Theta^2 = -1$ for spin $1/2$ electrons, Kramers' theorem applies, which leads
to the important constraint that each eigenstate of the time-reversal invariant
Hamiltonian has a partner of opposite k and opposite spin with the same
energy, the so-called Kramers' partner, which are at least two-fold degenerate.
However the strong SO coupling present in the 2D TI systems leads to a lifting
of the spin degeneracy of the states so that spin-up and spin-down states are
decoupled. Only at the time-reversal invariant points (Kramers' points), the
degeneracy persists due to Kramers' theorem and forms 2D Dirac points in
the band structure. The energy bands thus come in pairs and form Kramers'
partner (Fig. 2.6). In the Brillouin zone of a square lattice, there are 4 time-

---

[2]The following derivation of the time-reversal polarization as a topological index is based
on the original approach by Fu and Kane [12] and on the explanations given in ref. [56].

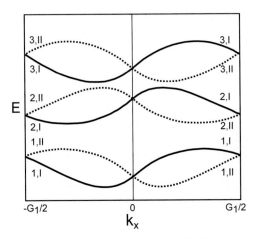

Figure 2.6: One dimensional band structure $E(k)$ with SO interaction along the k_x-direction as marked by the blue line (labeled c_1) in Fig. 2.5 b). The energy bands come in time-reversed pairs and cross at the time-reversal invariant points $\Gamma = 0$, $G_1/2$ (π/a_x). (Adopted from [12]).

reversal invariant points Γ_i with $\Gamma_i = n_i \, \mathbf{G}/2$, where $\mathbf{G}$ is a reciprocal lattice vector and $n_i = 0$ or 1 [12, 14, 13]. These points satisfy the relation:

$$H(\Gamma_i) = \Theta H(\Gamma_i)\Theta^{-1}. \tag{2.20}$$

As discussed above, the goal is to deduce an expression reflecting the change in the time-reversal polarization when k_y is moved from the center of the Brillouin zone to the boundary. The time-reversal polarization for each k_y thereby depends on the occupied states along the corresponding k_x-direction which projects onto the respective k_y-momentum (cf. Fig. 2.5 b)). The 1D band structure for the time-reversal system along the k_x-direction is shown in Fig. 2.6. The two bands which form a Kramers' pair (labeled as (n, I) and (n, II), respectively) are related to each other by a time reversal operation. The situation is such that the two bands I and II have opposite spin and move in opposite x-directions across the cylinder. Due to the connection of the states by Kramers' theorem, the $2N$ eigenstates of the system can be divided into N pairs which satisfy the following relations:

$$\left| u_n^I(-k_x) \right\rangle = -e^{iX_{k_x,n}}\Theta \left| u_n^{II}(k_x) \right\rangle$$
$$\left| u_n^{II}(-k_x) \right\rangle = e^{iX_{-k_x,n}}\Theta \left| u_n^I(k_x) \right\rangle, \tag{2.21}$$

The second relation can be derived from the first together with the property $\Theta^2 = -1$ which is valid for spin 1/2 electrons. Thus, time-reversal transforms eigenstates at k_x of bands I into eigenstates at $-k_x$ of bands II, and vice versa, but only up to an arbitrary phase $X_{k_x,n}$, which does not influence the Kramers' degeneracy. The importance of the topological analysis is to track this arbitrary phase, which is often neglected.

One can start the calculation by firstly determine the charge movement in x-direction, i.e. deducing an expression for the charge polarization P_x^I and P_x^{II} for each band I and II of a Kramers' pair, as the spin movement is equivalent to the movement of charge with different spin in different direction. Later, the expression of the time-reversal polarization will be directly derived from the partial charge polarizations P_x^I and P_x^{II}.

In general, the charge polarization of occupied bands may be deduced by defining Wannier functions for each set of occupied bands n and each lattice vector $\boldsymbol{R}$ in x-direction, so that the position of the occupied states in real space can be tracked [57]:

$$|\boldsymbol{R}, n\rangle = \frac{1}{2\pi} \int_{-G/2}^{G/2} dk_x \, e^{-ik_x(\boldsymbol{R}-x)} \, |u_n(k_x)\rangle . \tag{2.22}$$

As was shown by King-Smith and Vanderbilt [58], the charge polarization $P_x(k_y)$ in x-direction for each k_y, is then a function of the sum of the center of charge of the Wannier state at $\boldsymbol{R} = 0$, taken over all the bands n in k_x-direction:

$$P_x(k_y) = \sum_n \langle \boldsymbol{R} = 0, n| \, x \, |\boldsymbol{R} = 0, n\rangle = \frac{1}{2\pi} \int_{-\pi/a_x}^{\pi/a_x} dk_x \, i \sum_n \langle u_n^v(k_x)| \, \nabla_{k_x} \, |u_n^v(k_x)\rangle . \tag{2.23}$$

The first expression thereby describes if the center of mass of the charge of this state is shifted to the left or the right with respect to $R = 0$. Analogous to the description of the Chern number (section 2.2.2 and eq. 2.12) the integrand in eq. 2.23 is defined as the Berry vector potential with:

$$A(k_x) = i \sum_n \langle u_n(k_x)| \, \nabla_{k_x} \, |u_n(k_x)\rangle . \tag{2.24}$$

$P_x(k_y)$ as it is written in eq. 2.23 reflects the sum of the Wannier centers of all of the occupied bands. As in the described system the Wannier states come in Kramers' degenerate pairs, one may calculate the partial charge polarization for each sector I and II of all n Kramers' pairs individually. Analogously to eq. 2.24, one thus defines for the sector I (and the same for sector II):

$$A^I(k_x) = i \sum_n \left\langle u_n^I(k_x)\right| \nabla_{k_x} \left|u_n^I(k_x)\right\rangle , \tag{2.25}$$

and obtain for the partial charge polarization P_x^I:

$$
\begin{aligned}
P_x^I &= \frac{1}{2\pi} \int_0^{\pi/a_x} dk_x\, A^I(k_x) + \int_{-\pi/a_x}^0 dk_x\, A^I(k_x) \\
&= \frac{1}{2\pi} \int_0^{\pi/a_x} dk_x\, A^I(k_x) + \int_0^{\pi/a_x} dk_x\, A^I(-k_x).
\end{aligned}
\tag{2.26}
$$

The term $A^I(-k_x)$ can be rewritten to $A^I(-k_x) = A^{II}(k_x) - \sum_n \frac{\partial X_{k_x,n}}{\partial k_x}$ by using the time reversal constraint of eq. 2.21 and inserting it into eq. 2.25. Together with the relation $A(k_x) = A^I(k_x) + A^{II}(k_x)$, which is the sum of the Berry potential of sector I and II, one gets:

$$
P_x^I = \frac{1}{2\pi} \int_0^{\pi/a_x} dk_x\, A(k_x) - \sum_n (X_{k_x=\pi/a_x,n} - X_{k_x=0,n}).
\tag{2.27}
$$

Since the path of integration only covers half of the Brillouin zone the two remaining phases in eq. 2.27 may be different. However, the second term is necessary to preserve gauge invariance. This term can be rewritten by introducing a unitary matrix w_{mn} which relates the time reversed eigenstates at k_x with states at $-k_x$ from both sectors I and II:

$$
w_{mn}(k_x) := \langle u_m(-k_x) | \Theta | u_n(k_x) \rangle.
\tag{2.28}
$$

From the time reversal constraint of eq. 2.21 and the orthogonality of states, it follows that the only nonzero terms are $e^{iX_{k_x,n}}$ and $-e^{iX_{-k_x,n}}$ on the off-diagonal. Thus, the matrix does only contain the phase factors of degenerate pairs, which are the time-reversal invariant points $k_x = 0$ and $k_x = \pi/a_x$. Only at these points, w_{mn} is antisymmetric. At these points, one can define a so-called Pfaffian $Pf(w)$[3], which characterizes an antisymmetric matrix and whose square is equal to the determinant. The second term in eq. 2.27 may then be expressed in terms of $Pf(w)$:

$$
\frac{Pf[w(\pi/a_x)]}{Pf[w(0)]} = exp\left[i \sum_n (X_{k_x=\pi/a_x,n} - X_{k_x=0,n}) \right],
\tag{2.29}
$$

---

[3]The determinant of skew-symmetric matrix A (skew-symmetric matrix is a square matrix A whose transpose is its own negative, $A = -A^T$) can always be written as the square of a polynomial in the matrix entries, named the Pfaffian of the matrix $Pf(A)$: $det(A) = Pf(A)^2$. Introduced by [59], the Pfaffian is nonvanishing only for $2n \times 2n$ skew-symmetric matrices, in which case it is a polynomial of degree n (adopted from ref. [56]). In particular, if $A = \{a_{i,j}\}$ is the $2n \times 2n$ skew-symmetric matrix, then the Pfaffian of A is defined by:

$$
Pf(A) = \frac{1}{2^n n!} \sum_\sigma sgn(\sigma) \prod_{i=1}^n a_{\sigma(2i-1),\sigma(2i)}.
$$

and thus,

$$P_x^I = \frac{1}{2\pi} \left[\int_0^{\pi/a_x} dk_x \, A(k_x) + i \ln \frac{\mathrm{Pf}[w(\pi/a_x)]}{\mathrm{Pf}[w(0)]} \right]. \qquad (2.30)$$

The same calculation can be performed for P_x^{II} using the same arguments. Hence, the total charge polarization for a specific k_y is the sum of the charge polarization from both sectors I and II, $P_x^{\mathrm{total}} = P_x^I + P_x^{II}$. The time-reversal polarization, i.e. the relative polarization of the time-reversal partners, however is defined as:

$$P_x^{\mathrm{TR}} = P_x^I - P_x^{II} = 2P_x^I - P_x^{\mathrm{total}} \qquad (2.31)$$

In the following, it comes out that the relative polarization of time-reversal partners P_x^{TR} is either one or zero. This goes back to the fact that P_x^{TR} is expressed in terms of Pfaffians which are related to the determinant by $\mathrm{Pf}(A)^2 = \det(A)$. This leaves two possible signs of the Pfaffians with respect to the determinant, which later determine P_x^{TR} to be one or zero.

From eq. 2.31, the time-reversal polarization P_x^{TR} can be expressed in the form:

$$P_x^{\mathrm{TR}} = \frac{1}{2\pi} \left[\int_0^{\pi/a_x} dk_x \, A(k_x) - \int_{-\pi/a_x}^0 dk_x \, A(k_x) + 2i \ln \frac{\mathrm{Pf}[w(\pi/a_x)]}{\mathrm{Pf}[w(0)]} \right]. \qquad (2.32)$$

This expression may be written in a more compact way by using the trace Tr of the matrix w_{mn}, which relates the Berry vector potentials $A(k_x)$ at the momenta k_x and $-k_x$ with the unitary matrix w_{mn} in the following way (for a detailed derivation, see ref. [60] and [57]):

$$\mathrm{Tr}\left[w^T \nabla_k w\right] = (A(-k_x) - A(k_x))/i, \qquad (2.33)$$

so that one obtains:

$$P_x^{\mathrm{TR}} = \frac{1}{2\pi i} \left[\int_0^{\pi/a_x} dk_x \, \mathrm{Tr}\left[w^T \nabla_k w\right] - 2 \ln \frac{\mathrm{Pf}[w(\pi/a_x)]}{\mathrm{Pf}[w(0)]} \right]. \qquad (2.34)$$

Using the unitary property of w_{mn}, the first term can be written as:

$$\mathrm{Tr}\left[w^T \nabla_k w\right] = \mathrm{Tr}\left[\nabla_k \ln w(k_x)\right] = \nabla_k \ln \det\left[w(k_x)\right], \qquad (2.35)$$

and P_x^{TR} expressed as:

$$\begin{aligned}
P_x^{\mathrm{TR}} &= \frac{1}{2\pi i} \left[\ln \frac{\det[w(\pi/a_x)]}{\det[w(0)]} - 2 \ln \frac{\mathrm{Pf}[w(\pi/a_x)]}{\mathrm{Pf}[w(0)]} \right] \\
&= \frac{1}{\pi i} \ln \left(\frac{\pm\sqrt{\det[w(\pi/a_x)]}}{\pm\sqrt{\det[w(0)]}} \cdot \frac{\mathrm{Pf}[w(0)]}{\mathrm{Pf}[w(\pi/a_x)]} \right).
\end{aligned} \qquad (2.36)$$

Hence, one ends up with an expression for the time-reversal polarization P_x^{TR}, which depends on the Pfaffians and determinants at the specific points $k_x = 0$ and $k_x = \pi/a_x$ in the corresponding Brillouin zone. These factors only include the phase factors of the Kramers' pairs at the two time-reversal invariant momenta (TRIM) in one direction of the Brillouin zone. Since $\text{Pf}(A)^2 = \det(A)$ and due to the ambiguity of the logarithm of complex numbers by 2π, it follows that P_x^{TR} is an integer which is only defined modulo 2, thus is either 0 or 1. This means that a gauge transformation only changes the value of P_x^{TR} by an even integer, such that even or odd values of P_x^{TR} will go to even or odd values of P_x^{TR} under a gauge transformation. Whether the time-reversal polarization P_x^{TR} of the system has an even or odd value depends on the branches of the square root (+ or -) of the $\det(A)$. If the branches are chosen to be continuous along the way from $k_x = 0$ to $k_x = \pi/a_x$, the ambiguity of the square root is eliminated and the only question remains, whether the two fractions of determinant and Pfaffian $\left(\dfrac{\sqrt{\det[w(k_x)]}}{\text{Pf}[w(k_x)]} \right)$ are the same at $k_x = 0$ and $k_x = \pi/a_x$ or not. So, either the fractions have the same sign at both points or they have a different sign at both points, leading to an even or odd value for the time-reversal polarization.

This basic considerations of two possible outcomes for P_x^{TR} can be reformulated into the following expression:

$$(-1)^{P_x^{TR}} = \frac{\sqrt{\det[w(0)]}}{\text{Pf}[w(0)]} \frac{\sqrt{\det[w(\pi/a_x)]}}{\text{Pf}[w(\pi/a_x)]}, \qquad (2.37)$$

which is then either zero or one. The question, if there is a time-reversal polarization in opposite directions for the two Kramers' partners can therefore be answered by yes ($P_x^{TR} = 1$) or no ($P_x^{TR} = 0$). However, it turns out that the value of P_x^{TR} is not meaningful by itself as it depends on the chosen gauge of the wave functions, so a gauge transformation changes its value. The change in the time-reversal polarization, however, from $k_y = 0$ to $k_y = \pi/a_y$ is well defined and leads to a topological classification of specific pumping processes. In Fig. 2.5, the notion of a complete pumping cycle in terms of the introduction of exactly one flux quantum through a well defined cylinder was introduced, corresponding to the movement of momentum k_y from 0 to $2\pi/a_y$. One can now argue that the change in the time-reversal polarization which occurs in only *half* the cycle (from $k_y = 0$ to π/a_y) defines the topological invariant for a quantum spin Hall system.

Figure 2.7 a) provides a physical picture of the topological invariant by considering the shift of the Wannier state centers in the course of one cycle. At $k_y = 0$ and $k_y = \pi/a_y$, time-reversal symmetry requires that the Wannier states come in pairs, i.e. that the center of the occupied Wannier partners are

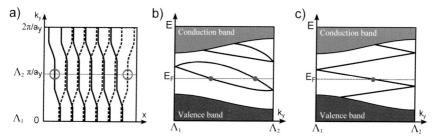

Figure 2.7: a) Sketch reflecting the position of the occupied Wannier states after their separation into two Kramers' partner (solid and dashed lines) going between two time-reversal invariant points $k_y = 0$ and $k_y = \pi/a_y$. The partners move in opposite x-direction accompanied with a change in the time-reversal polarization, which is a sign that the Wannier states switch partners (Adopted from [12]). b) and c) One dimensional $E(k)$ band structures along the edge of a 2D ribbon of finite width and infinite length. The electronic dispersion is shown between two Kramers' degenerate points $\Lambda_1 = 0$ and $\Lambda_2 = \pi/a_y$. The shaded areas mark the projected bulk valence and conduction bands of the 2D bulk, respectively. The edge states connect the Kramers' points in two different ways, shown in (a) and (b), reflecting the change in time-reversal polarization ΔP^{TR} between those points. Case (c) occurs in topological insulators and guarantees a necessary edge band crossing of the Fermi level E_F inside the bulk band gap. In (b), the edge states only cover parts of the bulk band gap and thus can not be topological (Adopted from [9]).

equal. However as the states of each pair move in opposite x-direction between the two TRIMs, the Wannier states switch partners. In this course, the time-reversal polarization is switched by one as it accounts for the difference between the positions of the Wannier states. As a consequence, the switching results in the appearance of an unpaired occupied Wannier state at $k_y = \pi/a_y$ for each end of the square (marked by circles in Fig. 2.7 a)), whereas in the bulk only a relabeling of the Wannier states take place. If the cycle moves on from $k_y = \pi/a_y$ to $k_y = 2\pi/a_y$, one ends up with identical wave functions as at the beginning. However the movement of the Wannier states in this second half of the cycle is the same as in the first half of the cycle, as can be shown by explicit calculation, so that a complete cycle moves all spins by two lattice constants in opposite directions. Consequently, the system returns to its former time-reversal polarization defined modulo 2, i.e. the time-reversal polarization at $k_y = 0$ and $k_y = 2\pi/a_y$ is equal. Hence, the change in time-reversal polarization which occurs due to the switching of Wannier partners and which defines the $\mathbb{Z}_2$ topological invariant occurs on half the cycle between $k_y = 0$ and $k_y = \pi/a_y$.

Based on eq. 2.37 and together with the above considerations one can now formulate an expression for the $\mathbb{Z}_2$ topological invariant in a 2D system. The 2D Brillouin zone with the four bulk TRIMs Γ_{ij} of a 2D system was already shown in Fig. 2.5 b). As discussed, two bulk time-reversal momenta at $k_x = 0$ and π/a_x (Γ_{11} and Γ_{12}) may be projected into one point denoted by $k_y = 0$ (Λ_1) on the edge, while the other two bulk TRIMs at $k_x = 0$ and π/a_x (Γ_{21} and Γ_{22}) are projected onto one point on the edge denoted by $k_y = \pi/a_y$ (Λ_2). The difference between the time-reversal polarization at these two edge TRIMs finally defines the topological invariant. Using eq. 2.37, one may write for each edge TRIM point Λ_i:

$$
\begin{aligned}
(-1)^{P_{\Lambda_i}^{\mathrm{TR}}} &= \frac{\sqrt{\det[w(\Gamma_{i1})]}}{\mathrm{Pf}[w(\Gamma_{i1})]} \frac{\sqrt{\det[w(\Gamma_{i2})]}}{\mathrm{Pf}[w(\Gamma_{i2})]} \\
&\equiv \delta_{i1} \cdot \delta_{i2} \\
&\equiv \pi_i = \pm 1.
\end{aligned}
\tag{2.38}
$$

Taking into account that the change in time-reversal polarization from Λ_1 to Λ_2 is gauge invariant, the topological invariant for a 2D system is:

$$
\nu = \left[P_{\Lambda_2}^{\mathrm{TR}} - P_{\Lambda_1}^{\mathrm{TR}} \right] \mathrm{mod}2,
\tag{2.39}
$$

and finally:

$$
\begin{aligned}
(-1)^{\nu} &= \prod_{i,j=1}^{2} \frac{\sqrt{\det[w(\Gamma_{i,j})]}}{\mathrm{Pf}[w(\Gamma_{i,j})]} \\
&= \prod_{i,j=1}^{2} \delta_{ij}.
\end{aligned}
\tag{2.40}
$$

Since $\det(w(\Gamma_{i,j})) = \mathrm{Pf}(w(\Gamma_{i,j}))^2$, the right-hand side of eq. 2.40 is always +1 or -1. Consequently, $\nu = 0$ or 1, with $\nu = 0$ denoting no change in the time-reversal polarization from Λ_1 to Λ_2 and thus topologically trivial and $\nu = 1$ denoting a change in the time-reversal polarization and thus topologically non-trivial.

The major fingerprint of a non-trivial system is the spin-polarized, or more precisely the time-reversal polarized edge states which cross the whole bulk band gap, accounting for the spin transfer present in the system. This crossing is a direct consequence of the change of the time-reversal polarization between the two TRIMs. The corresponding 1D band structure for the non-trivial case is depicted in Fig. 2.7 c). The edge states switch partners at the TRIMs and cover the whole band gap. As a consequence, the Fermi level E_F intersects the edge state only an odd times whereas in the case of a trivial insulator, E_F cuts the edge states an even times (Fig. 2.7 b)). Moreover in the trivial case, there can always be an energy region present within the bulk band gap at which E_F is not intersecting any edge states. The system is thus gapped.

The above calculations are all based on occupied wave functions of the bulk only, such that the edge states are a direct consequence of some bulk property. Since this property is not changed by moving E_F within the bulk gap, there have to be edge states at each energy within the gap. Further, the construction of the $\mathbb{Z}_2$ invariant ν relies on time-reversal invariance. One can show that the k-space representation is not required for the derivation. This implies that the edge states in the quantum spin Hall insulator are robust against the effects of interactions, even when spin conservation is broken [42, 8], as long as time-reversal symmetry is intact in the system and the character of the bulk band gap is not changed.

2.2.5 Generalization to three-dimensions

In this section, the approach of $\mathbb{Z}_2$ topological indices is applied to three-dimension, just by replacing the four time-reversal momenta for 2D squares by the eight TRIMs for 3D cubes. In general, in a 3D system, the TRIMs can be expressed in terms of primitive reciprocal lattice vectors as $\Gamma_{i=(n_1 n_2 n_3)} = (n_1 b_1 + n_2 b_2 + n_3 b_3)/2$, with $n_j = 0, 1$ [13]. In the case of a rectangular lattice, for example, the 8 TRIMS then correspond to the corners of the irreducible representation of the 3D Brillouin zone. For the sake of simplicity, the discussion is restricted to this rectangular unit cell. Analogous to the 2D case, where the 4 bulk TRIMs have been projected to two edge momenta in order to determine the time-reversal polarization of the two time-reversal invariant momenta in k_y-direction, one can now consider a slab with two surfaces cut perpendicular to the reciprocal lattice vector $\mathbf{G}$. Then, four time-reversal invariant points Λ_i remain. Their time-reversal polarization can be calculated by projections of pairs of Γ_i perpendicular to the slab which differ by $\mathbf{G}/2$ (see Fig. 2.8). One can again attribute a $\delta_{i=(n_1 n_2 n_3)} = \frac{\sqrt{\det[w(\Gamma_i)]}}{\mathrm{Pf}[w(\Gamma_i)]} = \pm 1$ to each of the 8 TRIMs of the 3D bulk. From its sign, the time-reversal polarization of each surface TRIM Λ_i can be determined by multiplying the related two Γ_i. The time-reversal polarization associated to the surface momentum Λ_a is then expressed by (similar to eq. 2.38 in the 2D case):

$$\begin{aligned}
\pi_a &= \delta_{a1} \cdot \delta_{a2} \\
&= \pm 1,
\end{aligned} \tag{2.41}$$

where δ_{a1} and δ_{a2} correspond to the bulk TRIMs which are projected onto the surface TRIM Λ_a. The sign of π_a for all a is not gauge invariant, however, the

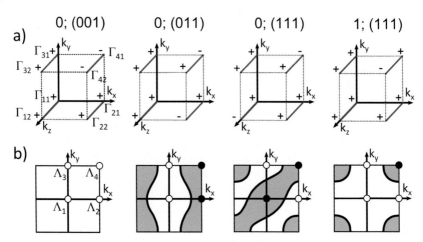

Figure 2.8: a) Sketches of four different phases indexed by $\nu_0;(\nu_1\nu_2\nu_3)$. a) 3D Brillouin zone with signs of δ_i marked at the corners which represent the 8 TRIMs Γ_i. Pairs of TRIMs used for the projection in (b) are connected by thick red lines. b) Projected surface Brillouin zone in the k_z direction [(001)-surface] of each phase. The projected surface TRIMs Λ_a are marked by open (closed) circles, depending on the resulting time-reversal polarization deduced by the 2 corresponding TRIMs Γ_i of the projected line. An open (closed) circle for Λ_a thus marks a +(-) for the product of the δ_i connected by a thick red line ($\pi_a = \delta_{a1} \cdot \delta_{a2} = +1(-1)$). Thick black lines separating the green from the white areas mark the Fermi lines of the surface state, which have to be present at each path between a white and a black dot, as this path represents a change in the time-reversal polarization requiring a surface state to be present at E_F. (Adopted from [13]).

product of $\pi_a\pi_b\pi_c\pi_d$ which determines the change in time-reversal polarization between the four TRIMs of a distinct slab direction is gauge invariant. As in the 2D case, the change in the time-reversal polarization determines a $\mathbb{Z}_2$ invariant which can be attributed to a specific surface direction. The corresponding bulk band structure of the system, normal to the surface in question, necessarily has the same constraints as in the 2D case. Namely that on any path between two TRIMs, which have opposite sign for π_a there must be a spin-polarized surface state crossing E_F. If only one of the four surface TRIMs has a (-) sign, the surface state forms a Fermi arc around this TRIM. Completion of the surface state in k-space and time-reversal invariance require, that the opposite spin components meet at the TRIM, which, thus, form a Dirac point surrounded by Dirac cones.

Now, a $\mathbb{Z}_2$ invariant can be attributed to each possible surface of the 3D system which accounts for the different combinations of δ_i, however, due to the crystal symmetry, only three of the invariants are independent [12, 49]. Moreover, there is a more general topological invariant, which tells, if all possible surfaces necessarily have surface states. This is obvious for the case of Fig. 2.8 a), labeled 1;(111), where only one of the eight TRIMs has $\delta_i = -1$, while the other seven have $\delta_i = +1$, such that always one pair of TRIMs gives a minus sign while the other pairs give a +sign independent of the direction of the surface. Consequently, a 3D system can be explicitly described by exactly four $\mathbb{Z}_2$ invariants which are nominated as $v_0;(v_1v_2v_3)$ [13, 14, 49, 54]. The first invariant v_0 is expressed as the product of δ_i of all eight TRIMs:

$$(-1)^{v_0} = \prod_{i=1}^{8} \delta_i. \tag{2.42}$$

The other three invariants are determined by the product of the four δ_i's for which the TRIMs Γ_i lay in the same plane:

$$(-1)^{v_v} = \prod_{n_k=1,n_{j\neq k}=0,1} \delta_{n_1 n_2 n_3}. \tag{2.43}$$

$(v_1v_2v_3)$ are clearly not independent of primitive lattice vectors $\mathbf{b}_k$. $(v_1v_2v_3)$ can be identified with a reciprocal lattice vector $\mathbf{G}_v = v_1\mathbf{b}_1 + v_2\mathbf{b}_2 + v_3\mathbf{b}_3$ and interpreted as Miller indices for $\mathbf{G}_v$.

Figure. 2.8 shows four distinct combinations of the $\mathbb{Z}_2$ invariants $v_0;(v_1v_2v_3)$ with the corresponding signs of δ_i at the 8 TRIMs of the 3D Brillouin zone (Fig. 2.8 a)) along with the expected surface state spectrum for the (001)-surface (Fig. 2.8 b)). Each TRIM Λ_a in the surface Brillouin zone corresponds to the projection of two TRIMs Γ_i from the bulk Brillouin zone connected by the red line. First one takes a look at eq. 2.42, which gives the value for the invariant v_0. If only one sign of δ_i is different from the other seven ones (Fig 2.8, most right), eq. 2.42 provides $v_0 = 1$ and all three linearly independent planes have exactly one Fermi arc enclosing a surface TRIM Λ_a. As described above, this follows from the fact, that every possible surface plane which is built by a combination of parallel planes including bulk TRIMs Γ_i always has one TRIM Λ_a with a different sign of $\pi_a = \delta_{a1} \cdot \delta_{a2}$ than the other (one closed circle and three open circles, or vice versa). Consequently, the time-reversal polarization is changed on any path between two distinct points and there must necessarily be surface states crossing E_F on any possible surface. The smallest size of the Fermi line is obviously a point located at the TRIM. This point is called the Dirac point of the topological surface state. Thus, in the case of $v_0 = 1$, every surface plane of the bulk insulator has surface states connecting the bulk valence and conduction band. The surfaces are thus conducting, if E_F lies within the band gap. These materials are called *strong* 3D TIs. As the name already suggest, the topological surface states of a strong 3D

TI are robust against all time-reversal invariant perturbations and can only be destroyed if time-reversal symmetry is broken (e.g. in proximity of an external magnetic field or magnetic impurities) or if the band gap is closed and, thereby, the system changes its topological invariants. Note, that in crystals with inversion symmetry, the sign of δ_i is related to the inversion symmetry of the occupied wave functions at Γ_i [14]. Each occupied state gets either a factor -1 or +1 after the inversion and the product of these numbers defines δ_i.[4] This implies that, if one changes the parities of the occupied bands at a particular TRIM, for example by exchanging an occupied p-type band with an s-type band, one may change the sign of δ_i and thus changing the system from non-trivial to trivial. For the inversion of bands, however, one has to close the bulk band gap. Thus, the topological surface states are robust with respect to changes of the Hamiltonian including disorder, which do not close the gap. On the other hand, one can use this knowledge in order to create a topological phase, namely by inverting the p- and s-type bands around E_F only at one TRIM and leaving the normal order at any other TRIM, such that only one δ_i differs in sign from all the others. Later we will see, that this comprehension led to the theoretical discovery of many materials with inversion symmetry to be strong 3D TIs. As already mentioned above, in a simple representation of a strong 3D TI, the Fermi surface encloses only one Kramers' degenerate Dirac point on the 2D surface. Close to the Dirac point E_D, the surface of the system can be described in terms of a Dirac Hamiltonian with linear spectrum [9]:

$$H_{\text{surf}}(k) = v_F k \cdot \sigma, \tag{2.44}$$

where v_F describes the Fermi velocity and σ the spin of the electron. Due to time-reversal symmetry, the spin is locked to the momentum and thus rotates around the Fermi surface (Fig. 2.9 a)). The corresponding $E(k)$ dispersion results in a Dirac cone, which is thus formed by a spin-polarized surface state (Fig. 2.9 b)).

Beside the class of strong TIs, which reside from a specific combination of δ_i's, other combinations of δ_i's lead to different topological phases. These combinations are depicted in the first three sketches of Fig. 2.8. Here, the sign of two δ_i's differ from the other six, so that the topological invariant v_0 = 0. In any of these cases, a plane can be found where the corresponding surface TRIMs Λ_a all have the same sign for $\pi_a = \delta_{a1} \cdot \delta_{a2}$. These planes thus have no change in the time-reversal polarization between TRIMs and therefor no topologically protected surface state crossing E_F. These are the planes (001), (011) and (111) in the three first examples of Fig. 2.8 (from left to

---

[4]The parity numbers of the occupied wave functions are easily achievable in a theoretical calculation, so that the determination of the $\mathbb{Z}_2$ invariants in the case of an inversion symmetric crystal is highly simplified.

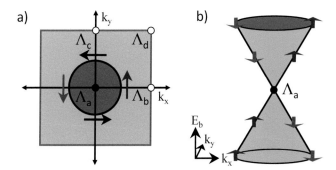

Figure 2.9: a) Surface Brillouin zone of a strong 3D TI with inverted bands at Λ_a such that a circular Fermi arc (marked by thick line) is formed around the TRIM (occupied states are marked in green). Arrows mark a possible spin direction of the states along the Fermi line. b) Corresponding $E(k)$ dispersion showing a Dirac cone which is formed by the spin-polarized edge states. The spin-polarized surface Dirac cone is the fingerprint of a strong 3D TI. (Adopted from [8]).

right). However, there are other planes which have two TRIMs with opposite sign, e.g. the (001)-surface in the second example of Fig. 2.8 (two open and two closed circled). Consequently, there are surface states present crossing the whole band gap but enclosing exactly two Dirac points. Materials which only have topological surfaces states on distinct surfaces are called *weak* 3D TIs. The term of weak 3D TIs (WTI) here results from the fact, that they always have an even number of Dirac points on their non-trivial surfaces, whereas in the case of a strong 3D TI (STI), there is always an odd number of Dirac points. Fu and Kane [14, 13] argued that in the case of disorder, it is likely that the two Dirac cones surrounding the Dirac points couple to each other and form a symmetric and antisymmetric combination of the wave functions. This might result in the opening of a gap within the topological surface states. The WTI would then be equivalent to a trivial band insulator, whereas the STI remains robust. However, in recent years it has been shown, that also the WTI has rather robust conductance with respect to disorder (more detail in the next chapter).

Regarding the consideration above, one is now able to explicitly classify the different topological phases by the four topological invariants $\nu_0;(\nu_1\nu_2\nu_3)$. If the index $\nu_0 = 1$, then the system is a STI with an odd number of conductive surface states on any of their surfaces. However, if $\nu_0 = 0$, the system is a trivial insulator or a WTI depending on the other three invariants $(\nu_1\nu_2\nu_3)$. If they are all zero, one has a trivial insulator without any topologically protected surface states at any of the surfaces. However, if $(\nu_1\nu_2\nu_3)$ are non-zero, then the triple marks the surface normal of the only surface without robust

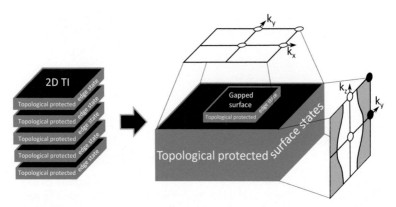

Figure 2.10: Schematic of a 3D weak topological insulator (right hand side) built by a stack of 2D TIs with the $\mathbb{Z}_2$ topological classification of 0;(001). The top surface of the WTI is gapped whereas the surrounding surfaces share topological protected surface states. A monolayer high island on the top surface is sketched, facing a 1D edge state at the step edge. The corresponding surface Brillouin zones for the top and the side surfaces are marked.

conductivity. The other surfaces have an even number of topologically protected surface states and the system is called a WTI.

2.2.6 Weak topological insulators

Weak topological insulators have above been described as a 3D system with an even number of topological surface states on only distinct surfaces of the crystal. Moreover, they have been reported to be "weak" with respect to disorder in a sense that a perturbation might gap the surface states. However in recent years, there have been several theoretical works claiming that the surface states are highly robust with respect to any time-reversal invariant perturbation [15, 16, 17, 18, 19, 20, 21].

Figure 2.10 shows the simplest way to create a WTI, namely by the stacking of 2D TI layers with topological protected edge states. A 3D system is formed, whereas the helical edge states of the layers then become anisotropic surface states (left side of Fig. 2.10). In this construction, the top and bottom surfaces (normal to the (001)-direction) of the WTI are the natural cleavage planes as well as the surfaces, which are topologically trivial. The four perpendicular surfaces, however, are topological non-trivial facing topological surface states which forms exactly two Dirac points. In the limit of completely decoupled layers, these surface states are the edge states of the stacked 2D TI. Translation-invariant coupling between the layers gaps out most of the surface along k_z, however, Kramers' theorem ensures the two Dirac points to remain [15]. The corresponding surface Brillouin zones for the top and side surfaces are marked in the right hand side of Fig. 2.10 revealing a possible surface Fermi distribution in the case of the side surface. This distinct WTI is thus classified by the $\mathbb{Z}_2$ invariant of 0;(001).

As already mentioned above, the major reason why the WTI is considered weak is, that the surface states might be destroyed without violating time-reversal symmetry or closing the bulk gap. However, the first belief that any type of random disorder would lead to a coupling of the two Dirac points and thus to a gapping of the surface states everywhere in 2D k-space turned out to be wrong. But, it was found that the way of coupling between the stacked layers in a WTI plays an important role with respect to the presence and robustness of the surface states. If one couples the layers into pairs, i.e. if the layers are dimerized, a mass term appears which opens up a gap at the Dirac point [15]. Hence, in the 3D limit of an even number of layers, all the surfaces turned out to be generally insulating if $E_D = E_F$. Whereas if the number of layers in a WTI is odd, there is no way to gap all the edge states only by pairing of the layers alone, and thus the surface remains conductive. As the dimerization of layers breaks the lattice-translation symmetry, it seems as this symmetry is essential for the protection of the topological phase in a WTI [15, 21].

However, disorder also breaks translation symmetry and one could think that this perturbation might lead to an Anderson localization of the surface states in a WTI in the presence of disorder as in a conventional metal [43]. Hence, Ringel *et al.* [15] showed in their work, using both a topological approach as well as a more quantitative perturbation analysis, that a non-trivial surface of a WTI remains conductive, namely that the longitudinal conductance σ_{xx} remains higher than e^2/h in the presence of disorder of arbitrary strength. Thus, only a pairing of the 2D TI layers, which leads to a distinct interaction of an edge state of a particular layer to a layer above than to a layer below, together with an even number of layers results in a gapping of the surface states at the non-trivial surfaces of a WTI [17, 16, 15]. Moreover, if the disorder in the coupling to adjacent layers is larger than the preferential coupling to one of the layers, the topologically protected conductance is restored. Thus, an even number of 2D TI layers alone without any dimerization of the layers or with a dimerization which is weaker than the disorder, is not sufficient to break the conductance of the surface states. Hence, it turns out that the WTI phase has a similar robust nature than the STI phase.

A very promising feature of the WTI is that any dislocation or step edge present at the trivial surface (the so-called "dark" surface) of the WTI might host a perfectly conducting spin-helical 1D edge mode which exhibits all the advantages of a topological state, most notably a lack of back-scattering and localization. Imagine for example a monolayer high island on the top surface of the WTI (right hand side in Fig. 2.10), then a closed loop of a perfectly conducting edge state must form exactly at the step edge of the island [20].

In recent years, there have been a few theoretical proposals of potential materials to face the weak TI phase, but none of them has been realized experimentally [61, 62, 63, 64]. It was only in 2013 that Rasche *et al.* [32] successfully synthesized the first proposed weak TI, namely $Bi_{14}Rh_3I_9$. It consists of layers with honeycomb structure, being identified to be 2D TIs, which are stacked and only weakly coupled by separating them by trivial insulator sheets in order to form a weak 3D TI. So far, a direct experimental demonstration of the weak TI properties is lacking. Within this work, I will demonstrate for this particular material the presence of backscatter-free 1D electron channels at the step edges of the gapped top surface, identifying the system as a weak TI.

2.2.7 Experimental realizations of 3D topological insulators

After the theoretical prediction of this new class of quantum matters in three dimensions and the first experimental verification of the QSHE in HgTe quantum wells [7], the first real materials with STI properties have been discovered soon. The starting point was to look for inversion symmetric materials with a band inversion at only one TRIM in the Brillouin zone, preferentially at the Γ-point, i.e. the center of the Brillouin zone. In that sense, a strong SO coupling, which is present in heavy elements, is helpful as it pushes the bands towards an inversion. Moreover, the materials should be insulating in the bulk and thus have a bulk band gap. This requirement is preferentially present in alloys.

The first experimental discovery of a strong 3D TI material was the alloy $Bi_{1-x}Sb_x$ [65, 66], but the surface band structure turned out to be relatively complex with five topological surface states crossing E_F. Soon after, Zhang *et al.* [67] proposed the well-known thermoelectric materials Bi_2Se_3, Bi_2Te_3 and Sb_2Te_3 to exhibit a STI nature with only one single Dirac cone at the Γ-point. These materials are inversion symmetric, consist of heavy elements, and exhibit a bulk band gap of approximately 200 meV. Zhang *et al.* showed that for these three materials, the SO coupling is strong enough to induce a band inversion only at the Γ-point, such that a state with positive parity exchanges with a state of negative parity at E_F, leading to the necessary change of sign for δ_i at Γ [67]. The density functional theory (DFT) calculation of the surface band structure of Bi_2Se_3 is depicted in Fig. 2.11 a) and reveals the topological surface state at the Γ-point connecting the bulk valence band (BVB) with the bulk conduction band (BCB) and forming a Dirac cone with a Dirac point.

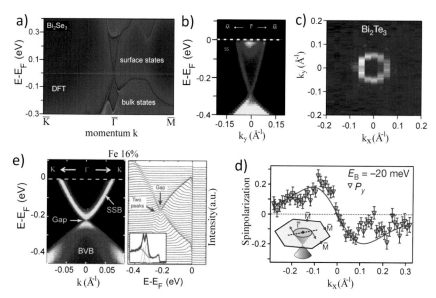

Figure 2.11: a) DFT calculation of the surface band structure of Bi_2Se_3 at the (111) cleavage plane. The warmer colors represent higher LDOS. Surface states and bulk bands are marked. (Taken from [67]). b) ARPES data of the same surface of Bi_2Se_3 along the ΓM-direction in k-space. (Taken from [68]). c) ARPES intensity map at E_F of the (111) surface of Bi_2Te_3. Red arrows mark the spin direction rotating around the Fermi surface. d) Measured y-component of the spin polarization at $E - E_F = -20$ meV along the ΓM-direction. The spin polarization inverts for opposite k-values as expected from time-reversal symmetry. The polarization is strongest in the in-plane direction perpendicular to k_x as visualized in the inset. ((c) and (d) are taken from [69]). e) Left and right panel show the ARPES data and the stacking plot of the energy distribution curves (EDCs) of a 16% Fe doped Bi_2Se_3 (111) sample, respectively. At the Dirac point, a reduced ARPES intensity (left image) and a twin-peak structure in the EDCs (right image) indicate a gap formation due to the breaking of time-reversal symmetry by the magnetic impurities. (Taken from [70]).

Experimentally, the presence of such a Dirac cone in the surface band structure of Bi_2Se_3, Bi_2Te_3 and Sb_2Te_3 has been verified by angle-resolved photoemission spectroscopy (ARPES)[5] in several works [68, 71, 72]. The ARPES spectrum of the (111) surface of Bi_2Se_3 is exemplarily shown in Fig. 2.11 b). It reveals the experimental surface Dirac cone located at the Γ-point in the center

---

[5]ARPES probes the band structure $E(\mathbf{k})$ of a particular surface. The angle of the emitted photoelectrons is thereby directly related to the k-vector parallel to the surface. For a detailed description of the technique, see chapter 3.2.

of the Brillouin zone. It further shows nice agreement with the DFT calculation (Fig. 2.11 a)) except that the Fermi level in the experiment is not located at the Dirac point but within the conduction band, indicating a strong n-doping within the material. The same tendency is observed for Bi_2Te_3, and oppositely for Sb_2Te_3, which is strongly p-doped and ascribed to vacancies and antisites defects produced during the growth process of the crystal [73, 74]. Thus, the predicted transport properties of the surface states are not expected for these compounds, unless the Fermi level is tuned into the gap regime (e.g. by Ca doping [69]).

Using spin-resolved ARPES, the spin information of the topological surface states is directly accessible and provides a profound argument for the topological nature of the surface states. Figure 2.11 c) shows the Fermi surface of the circular Dirac cone of Bi_2Te_3 together with the experimentally deduced spin orientation. The corresponding measured y-component of the spin polarization is depicted in Fig. 2.11 d) and reveals the expected spin inversion for opposite k-values. Hence, spin-resolved ARPES is a powerful tool in order to map the spin helicity of the topological surface states [69, 66, 75].

As already discussed in the sections above, the topological nature of the surface states remains intact as long as time-reversal symmetry is protected. However, if an impurity which breaks the time-reversal symmetry is introduced into the system, a gapping of the surface states is predicted. Experimentally, it has been shown that magnetic doping indeed opens up a gap within the crossing point of the Dirac cone [70, 76, 77]. Figure 2.11 e) shows an ARPES spectrum of a Bi_2Se_3 crystal with 16 % of Fe incorporated. At the Dirac point, a reduced ARPES intensity is visible. Together with the twin-peak structure found in the corresponding energy distribution curves (EDCs), the opening of a gap is obvious.

A different technique, which has proven its potential for the characterization of 3D TIs is scanning tunneling spectroscopy (STS)[6]. It records the differential conductivity dI/dV at a selected applied bias voltage between the probing tip and the sample, which is a direct measure of the local density of states (LDOS) of a particular region of the surface at a distinct energy. STS can thus resolve the electronic structure of the TI surface on an atomic scale. The electronic surface state properties can be made visible by scattering processes at impurities or step edges, which then result in a standing wave pattern of the surface states. A number of STS experiments has been done immediately after the experimental discovery of Bi_2Se_3, Bi_2Te_3 and Sb_2Te_3 as a 3D TI [80, 78, 81, 74, 82]. Figure 2.12 a) shows a Bi_2Te_3 (111) surface with adsorbed Ag trimers acting as scatters. The corresponding LDOS maps measured at various energies are depicted in c)-g) of Fig. 2.12. The changing intensity is

---

[6]For a detailed description of the technique see chapter 3.1.3.

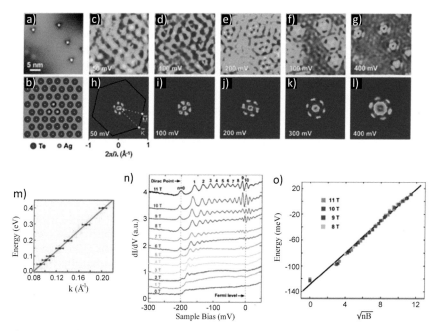

Figure 2.12: a) STM image of Ag trimers adsorbed on the (111) surface of Bi$_2$Te$_3$. b) Adsorption geometry of Ag trimer. c)-g) dI/dV maps of the same area as in a) at various sample bias voltages recording the LDOS as a function of energy. The interference pattern is caused by the overlapping of standing electron waves scattered at the Ag trimers. h)-l) Corresponding FFT spectra of the dI/dV maps in (c)-(g) revealing the possible scattering wave vectors **q** of the standing waves pattern. The surface Brillouin zone is exemplarily superimposed in (h) in order to indicate the directions of the **k**-space. m) $E(\mathbf{k})$ dispersion derived from the FFT spectra. ((a)-(m) taken from [78]). n) Local $dI/dV(V)$ spectra measured on the (111) surface of Bi$_2$Te$_3$, representing the LDOS as a function of energy at different B-fields as marked. The numbers n mark the consecutive Landau levels. o) Corresponding $E(\mathbf{k})$ dispersion deduced from the Landau levels. ((n) and (o) taken from [79]).

given by the standing waves of the electrons originating from the scattering at the Ag trimers. The Fourier Transformation (FT) of the dI/dV maps (Fig. 2.12 h)-l)) reveal the possible scattering vectors **q** between two states ($\mathbf{q} = \mathbf{k}_1 - \mathbf{k}_2$) in the system. By superimposing the surface Brillouin zone to the FFT spectra in Fig. 2.12 h), it gets obvious that the scattering vector in ΓK-direction is absent. As described in more detail in ref. [78], this particular scattering vector belongs to a direct backscattering between **k** and -**k** and is thus eliminated from the standing wave pattern by destructive interference, i.e. by the helical nature of the topological surface states. Hence, STS provides a direct confirmation of the lack of backscattering on the surface of a strong 3D TI.

The 2D nature as well as the cone structure of the topological surface states can further be demonstrated by STS, measuring the Landau level quantization on the surface of a 3D TI in a magnetic field [79, 83, 84]. Figure 2.12 n) and o) show the appearance of Landau level quantization in the $dI/dV(V)$ spectra at different magnetic field strength B as well as the corresponding $E(k)$ dispersion for the surface of a Bi_2Se_3 thin films [79]. It has been shown that the distance between Landau levels decreases with increasing energy from the Dirac point, which is located at the $n = 0$ Landau level at $E - E_F = $ -200 meV. Moreover, the Landau levels in this measurement obey the relation $E \propto \sqrt{nB}$ (Fig. 2.12 o)), which is valid for Dirac fermions and thus directly confirms the Dirac cone structure of the surface states.

Another STS work highlighting the special properties of the surface states of a topological insulator have been done by Seo *et al.* [85] measuring their transmission through atomic steps on a antimony (111) surface. In contrast to trivial surface states of common metals (e.g. copper, silver or gold), which are either fully reflected or absorbed by atomic steps, they found that the topological surface states of Sb possess a high probability ($\approx 35\%$ transmission) of penetrating such barriers. This again reflects an unique property of a topological surface state.

In the meantime, other interesting topological phases, which can be classified by topological indices have been discovered experimentally, such as topological crystalline insulators [86, 87] or the anomalous quantum Hall effect [88, 89]. As both being beyond the scope of this work, I will not describe these classes in more detail.

3 Experimental Methods

In the framework of this PhD thesis, predominately two surface sensitive techniques have been used in order to examine the atomic and electronic structure of different topological insulator systems. Scanning tunneling microscopy (STM), on the one hand, is a very powerful method regarding the imaging of the atomic arrangement of a conducting surface as well as its electronic properties on a sub-nanometer length scale. On the other hand, Angle-Resolved Photoemission Spectroscopy (ARPES) which also maps the electronic properties of a system, but on a larger length scale, depending on the spot size of the focused incident beam (10 μm to 1 mm). The big advantage with respect to STM, is the momentum-dependent acquisition of the electronic properties, rendering a complete picture of the band structure in k-space. This chapter briefly discusses the basic physical principals of both techniques, providing an overview of their particular strengths.

3.1 Scanning tunneling microscopy

STM has been developed in 1981 by Binnig and Rohrer [90]. In its operation mode, a sharp metallic tip is placed in front of a conducting surface in a distance, that a tunneling current occurs if a voltage between the tip and the sample is applied (mV to V) [91]. The distance between tip and sample is typically of the order of a few Angstrom (Å) an can be realized by various driving mechanisms based on piezo elements. The tunneling current thereby depends exponentially of the tip-sample distance which is the main reason for the extreme lateral and z-resolution, typically lying below the atomic length scale. The potential scheme of the tip-sample configuration shown in Fig. 3.1 gives a first explanation for the tunneling process in the case of a planar tunneling contact. The states in the tip and sample are filled up to the Fermi level E_F, and if a voltage between the tip and the sample is applied, electrons from the tip tunnel into empty states of the sample (positive sample bias in Fig. 3.1 a)) and vice versa (negative sample bias in Fig. 3.1 b)). In the approximation of a constant density of states for the tip and sample, the tunneling current can be described by:

$$I \propto V e^{-2\kappa d} \tag{3.1}$$

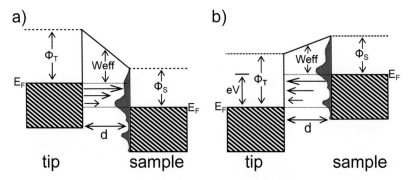

Figure 3.1: Potential model of a planar tunneling contact between a metallic tip and a sample with applied a) positive sample bias and b) negative sample bias. The dashed area marks the occupied states and the potential heights are given by the work function of the tip Φ_T and sample Φ_S, respectively. The Fermi energies E_F are shifted by the value of the applied bias voltage $e \cdot V$ and electrons from this energy range may tunnel into empty states. The tunneling probability depends on the energy of the electrons as well as on the density of states (shaded area).

with

$$\kappa = \sqrt{\frac{2m}{\hbar^2}\left(\Phi_{\text{eff}} - \frac{|eV|}{2}\right)}, \tag{3.2}$$

and Φ_{eff}, the average work function of the respective work functions of the tip Φ_T and the sample Φ_S, which is typically of the order of a few electron volts. In the case of a small bias voltage (mV regime), the relationship 3.1 between the current and bias is close to linear, but if the bias increases, the dependence of the exponent κ of the bias V gets dominant and there is a reduction of the effective tunneling barrier $W_{\text{eff}} = \Phi_{\text{eff}} - \frac{|eV|}{2}$, leading to a stronger increment of the tunneling current I with bias. Furthermore, I exponentially depends on the barrier width d, which already indicates the high spatial resolution of an atomically sharp STM tip in z-direction. As not taken into account here in this simple current-bias relationship, the tunneling current also depends on the density of states in the energy range given by the applied bias voltage $e \cdot V$. A three dimensional and more quantitative description of the tunneling process in an STM experiment will be discussed in the following section.

3.1.1 Tersoff-Hamann model

Based on the mathematical approach for the elastic tunneling between two metallic layers by Bardeen [93], J. Tershoff and D. R. Hamann [94, 92] developed a theory for the tunneling process in STM, which included an exact

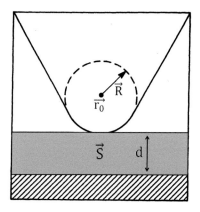

Figure 3.2: Sketch reflecting the tunneling geometry according to the Tersoff-Hamann model (after [92]). A spherical STM tip with radius R at a position $\vec{r}_0$ in a distance d to the sample (shaded area).

description of the tunneling current I in a 3D tunneling barrier. In the limit of small bias voltages with respect to the work-function Φ ($e \cdot V \ll \Phi$) and low temperature ($T \to 0\,\mathrm{K}$), for which the broadening of the Fermi distribution can be neglected, the tunneling current I can be written:

$$I \propto \int_0^{eV} |M_{TS}|^2 \rho_T(E_{F,T} - eV + \varepsilon)\rho_S(E_{F,S} + \varepsilon)d\varepsilon, \tag{3.3}$$

with $E_{F,T}$ the Fermi energy of the tip and $E_{F,S}$ the Fermi energy of the sample, ρ_T and ρ_S the density of states of the tip and sample at a certain energy and M_{TS} the tunneling matrix element between the tip and sample wavefunctions (Ψ_T and Ψ_S, respectively), which accounts for the transition probability between states of tip and sample. According to Bardeen [93], the tunneling matrix elements can be determined by:

$$M_{TS} = \frac{-\hbar^2}{2m} \cdot \int d\vec{S} \cdot (\Psi_T^* \vec{\nabla} \Psi_S - \Psi_S \vec{\nabla} \Psi_T^*), \tag{3.4}$$

where $\vec{S}$ reflects an area within the tunneling barrier (see Fig. 3.2). As the exact crystal structure of the endmost atoms of the tip is not accessible due to spontaneous rearrangements, the wave-function Ψ_T can not be determined accurately. Tersoff and Hamann therefore assume a spherical s-type wavefunction for the tip which then allows the determination of the tunneling matrix elements. The tunneling current I from eq. 3.3 may then be simplified to [94, 92]:

$$I \propto Ve^{2\kappa R}\rho_T(E_{F,T})\rho_S(E_{F,S}, \vec{r}_0), \tag{3.5}$$

where one assumes a constant density of states of the tip ρ_T at $E_{F,T}$. $\rho_S(E_{F,S}, \vec{r}_0)$ is the local density of states of the sample at $E_{F,S}$ at the center position of the s-type wave-function representing the tip $\vec{r}_0$ (see Fig. 3.2). The decay constant κ is given by $\kappa = \sqrt{\frac{2m_e\Phi_{\text{eff}}}{\hbar}}$, with m_e being the electron mass and Φ_{eff} the effective work function, described in first approximation by $\Phi_{\text{eff}} = \frac{\Phi_T + \Phi_S}{2}$.

Since most STM measurements are done with W- or PtIr-tips, which typically do not have s-type wave-functions but mostly spatially extended d-type orbitals, the resolution in the experiment can drastically increase. C. J. Chen [95] therefor extended the Tersoff-Hamann-model for other tip orbitals deriving the so-called derivation rule, i.e. Ψ_S has to be replaced by $\frac{d\Psi_S}{dz}$ for p_z-type tip orbitals, by $\frac{d^2\Psi_S}{dz^2}$ for d_{z^2}-type orbitals, by $\frac{d\Psi_S}{dx}$ for p_x-type tip orbitals and so forth. Notice, that only the high angular momentum orbitals at the tip (i.e. p, d,...) are able to explain the lateral atomic resolution, which has been achieved on close-packed metal surfaces, such as Al(111) [96] and Au(111) [97]. The measured corrugation was thereby much higher than predicted from the Tersoff-Hamann-model.

3.1.2 Scanning tunneling spectroscopy

As already stressed in the beginning of this chapter, STM is a very powerful tool for spectroscopic measurements, recording the electronic structure of a surface on a very local scale (sub-nm). As a large part of this work has been carried out by scanning tunneling spectroscopy (STS), the physical principles of this mode will be described in this section. When doing STS, one is usually interested in the density of states of a surface at a very distinct position $\rho_S(\vec{r}, E)$. Based on the description of the tunneling current I from the Tersoff-Hamann model, I can be simplified according to Selloni et al. [98] with $\varepsilon = E - E_F$ to:

$$I \propto \int_0^{eV} \rho_T(E_F - eV + \varepsilon) \cdot \rho_S(\vec{r}, E_F + \varepsilon) \cdot T(\varepsilon, V, s)d\varepsilon. \qquad (3.6)$$

$T(\varepsilon, V, s)$ is a transmission coefficient deduced from the simple model of two planar tunneling contacts based from the semi-classical WKB[1]-theory:

$$T \approx e^{-2\kappa(\varepsilon, V)s} \qquad (3.7)$$

with

$$\kappa(\varepsilon, V) = \sqrt{\frac{2m}{\hbar^2}\left[\Phi_{\text{eff}} + \frac{eV}{2} - \left(\varepsilon - \frac{\hbar^2 k_\parallel^2}{2m}\right)\right]}. \qquad (3.8)$$

---

[1]Wentzel-Kramers-Brillouin

Here, $s = d + R$ is the distance between the surface and the center of the tip s-orbital (see Fig 3.2) and $\Phi_{\mathrm{eff}} = \frac{\Phi_T + \Phi_S}{2}$ the averaged work function of the tip and sample. Considering the decay constant κ in more detail, it gets visible that κ depends on the applied bias voltage V and the band structure of the sample at the surface $\varepsilon(\boldsymbol{k})$, which is defined by the energy ε and the momentum $\boldsymbol{k}$ of the tunneling electrons. According to equation 3.7 the transmission T, and also the tunneling current I, increases if the decay constant κ gets small, which is the case for an electron of a certain energy ε with a vanishing k-momentum parallel to the surface ($k_{\parallel} = 0$). This means that states originating from the Γ-point, where $k_{\parallel} = 0$, contribute most to the tunneling current I in the STS measurements [99]. Subsequent derivation of eq. 3.6 then leads to:

$$
\begin{aligned}
\frac{dI}{dV}|_V \propto\ & e \cdot \rho_T(E_{F,T}) \cdot \rho_S(E_{F,S} + eV) \cdot T(eV, V, s) \\
& + \int_0^{eV} \rho_T(E_{F,T} - eV + \varepsilon) \cdot \rho_S(E_{F,S} + \varepsilon) \cdot \frac{d}{dV}[T(\varepsilon, V, s)] d\varepsilon \qquad (3.9) \\
& + \int_0^{eV} \frac{d}{dV}[\rho_T(E_{F,T} - eV + \varepsilon)] \cdot \rho_S(E_{F,S} + \varepsilon) \cdot T(\varepsilon, V, s) d\varepsilon.
\end{aligned}
$$

The first term in eq. 3.9 contains the local density of states of the sample at an energy of the applied bias voltage $e \cdot V$ with reference to the Fermi energy $E_{F,S}$. The second term is important in the case of higher bias voltage being about 10 % of the first term at $V = 200\,\mathrm{mV}$. For moderate bias voltage however ($V \ll \Phi_{\mathrm{eff}}$), which is the case for the STS measurements presented in this work, this term is negligible. The third term represents variations of the tip density of states with energy, and is negligible in case of $\frac{d\rho_T}{dV} \cdot V < \rho_T$. It can be reduced by trial and error preparation of the individual microtip. It follows that the acquisition of the differential conductivity dI/dV in the STS experiment provides nearly direct access to the local density of states (LDOS) of the sample surface at the applied bias voltage V.[1]

$$
\frac{dI}{dV}(V) \propto \rho_S(E_F + eV). \qquad (3.10)
$$

In the STS experiment, the differential conductivity dI/dV can be measured by lock-in technique. The applied bias voltage V is modulated by a sinusoidal modulation voltage V_{mod} (few mV) at a frequency f_{mod} of the order of 1-2 kHz. The tunneling current is thereby used as the input signal of the lock-in amplifier. The input signal is multiplied by the reference signal (V_{mod}) and

---

[1]Typically $\rho_S(E_{F,S} + eV)$ is the only factor of the first term of eq. 3.9 which depends on the position of the tip with respect to the sample, while, however, a spatial dependence of the vertical decay function $T(\varepsilon, V, s)$ cannot be excluded a priori.

integrated over a distinct time (time constant t_c). The output amplitude of the lock-in amplifier is then proportional to the differential conductivity dI/dV. This technique amplifies the small tunneling current variations during an STS experiment and leads to a better signal-to-noise ratio.

3.1.3 STM and STS experiments

In this work, topographic STM images down to the atomic scale have been recorded using the *constant current mode*. At a defined tunneling current I and bias voltage V, the tip of the STM scans an (x,y)-area of the sample surface while adjusting the tip surface distance s in vertical z-direction in order to keep I constant. The z-variation of the tip is pictured in a color plot $z(x,y)$ providing the so-called topography of a sample in the STM image, which more precisely is a spatial contour of constant local density of states (LDOS) integrated between the respective Fermi energies of tip $E_{F,T}$ and sample $E_{F,S}$.

STS measurements have been predominately used in order to gain local $dI/dV(V)$ curves (LDOS curves) or dI/dV images, which are spatially resolved images of the LDOS at a particular energy $E = e \cdot V$. The first method is an important tool in order to gain knowledge about variations of the LDOS with energy at a particular location. A $dI/dV(V)$ curve is measured by stabilizing the STM tip at a chosen position at the tunneling distance, defined by a stabilizing voltage V_{stab} and current I_{stab}. Then, a voltage ramp V_{ramp} is applied and the lock-in output collected, providing the differential conductivity with respect to the applied voltage ramp $(dI/dV(V))$. As described above, this value can be interpreted as the local density of states in a distinct energy range E, which is defined by the applied voltage ramp.

A dI/dV image, however, can be recorded by simultaneously measuring the output of the lock-in amplifier while recording an STM image at a distinct bias voltage and tunneling current. The applied bias voltage during the scanning defines the energy $(e \cdot V)$ at which the spatially resolved LDOS is measured and is abbreviated in the following by V_{stab}. This method is the fastest way to spatially resolve the LDOS of the surface but has the disadvantage that imaging in the energy regions of band gaps is not easy. Also measuring the LDOS at E_F ($V{=}0$) is not possible. In order to avoid this problem, a dI/dV map can be acquired. At each grid point of an STM image, a $dI/dV(V)$ curve is measured at a defined stabilizing voltage V_{stab}, providing a spatially resolved image of the LDOS as a function of energy $e \cdot V_{ramp}$.

Besides the high spatial resolution of the STM, the energy resolution plays an important role in the case of STS experiments. The energy resolution ΔE is given by [100]:

$$\Delta E \approx \sqrt{(3,3k_B T)^2 + (2,5eV_{mod})^2}, \tag{3.11}$$

where the temperature T and the modulation voltage V_{mod} are the main parameter. The limit of ΔE due to finite temperature is predominately the main reason of going to low temperatures in a STS experiment. In this work, the STS data has mostly been collected at a temperature of $T = 6\,\mathrm{K}$ and a modulation voltage of $V_{mod} = 4\,\mathrm{mV}$ (RMS), exhibiting an energy resolution of $\Delta E \approx 7\,\mathrm{meV}$.

3.2 Angle-resolved photoemission spectroscopy

Angle-resolved photoemission spectroscopy (ARPES) is the most direct method of studying the electronic structure of solids. It measures the surface electrons when emitted by an incident photon of a particular energy $h\nu$. Based on the escape angles and the kinetic energy of the photoelectrons, information of the electronic band structure $E(k)$ of the material is accessible. The angle of the emitted photoelectrons is hereby directly related to the k-vector within the surface. The geometry of an ARPES experiment is depicted in Fig. 3.3. It shows the incident photon which is a beam of monochromatic radiation of a certain energy $h\nu$, either supplied by a gas-discharge lamp or by a synchrotron beamline. The beam is focused on the sample and the electrons are emitted based on the photoelectric effect with a kinetic energy:

$$E_{kin} = h\nu - \Phi - E_{bin}, \qquad (3.12)$$

with Φ being the work function of the particular material (typically 4-5 eV in metals) and E_{bin} the binding energy of the electrons inside the crystal measured with respect to the Fermi level E_F. The emitted electrons are then collected by an electron energy analyzer characterized by a finite entrance slit, which selectively measures the electrons of a specific kinetic energy E_{kin}^{out} and a given emission direction. The emission angle is fully characterized by the polar angle Θ and azimuth angle φ and thus completely determines the momentum $\mathbf{k}^{out}$ of the photoelectron outside the crystal. Its magnitude is given by $k^{out} = \frac{1}{\hbar}\sqrt{2mE_{kin}^{out}}$. The component parallel and perpendicular to the surface $\mathbf{k}_\parallel^{out} = \mathbf{k}_x + \mathbf{k}_y$ and $\mathbf{k}_\perp^{out} = \mathbf{k}_z$, respectively, are obtained in terms of Θ and φ [101]:

$$k_x = \frac{1}{\hbar}\sqrt{2mE_{kin}^{out}}\,\sin\Theta\,\cos\varphi,$$

$$k_y = \frac{1}{\hbar}\sqrt{2mE_{kin}^{out}}\,\sin\Theta\,\sin\varphi, \qquad (3.13)$$

$$k_z = \frac{1}{\hbar}\sqrt{2mE_{kin}^{out}}\,\cos\Theta.$$

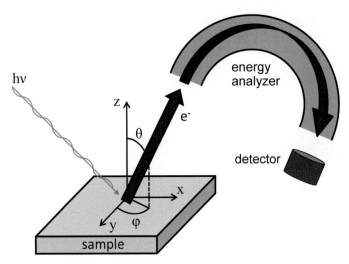

Figure 3.3: Geometry of an ARPES experiment. A photoelectron e^- is emitted from a sample by an incident photon of energy $h\nu$. The momentum k of the electron is encoded in its emission direction of the sample, given by the polar angle Θ and azimuth angle φ. A detector, which is situated behind an energy analyzer, captures the photoelectrons angle- and energy-dependent, which then gives rise to the $E(k)$ dispersion of the sample.

The goal is to deduce the desired electronic dispersion $E(k)$ of the solid, i.e. the relation between the binding energy E_{bin} and the momentum k of electron states inside the crystal, based on the measured kinetic energy E_{kin}^{out} and momentum k^{out} of the emitted photoelectrons outside the crystal. Taking advantage of the total energy conservation law in a non-interacting electron picture between an initial and final state ($E_f - E_i = h\nu$), the binding energy of a state inside the crystal can be deduced from the measured kinetic energy according to eq. 3.12:

$$E_{bin} = h\nu - \Phi - E_{kin}^{out}. \tag{3.14}$$

The energetic relation between E_{bin} and E_{kin}^{out} is visualized in the energy sketch of the photoemission process in Fig. 3.4 a).

However, gaining information of the full momentum k^{in} of the initial states is more complex. Here, one can take advantage of the so-called *three-step model* [102], which is sketched in Fig. 3.4 b), and in which the photoemission process is decomposed in three independent steps. It is the most common model used in the interpretation of photoemission data, in particular when photoemission spectroscopy is used as a tool to map the electronic band structure of solids. In the first step, the photoexcitation process is drawn as a tran-

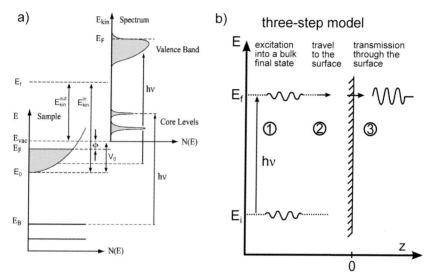

Figure 3.4: a) Energetic scheme of the photoemission process when an incoming photon emits electrons from the solid. On the bottom left side, the energy distribution of the crystal in terms of binding energy E_{bin} is depicted with the vacuum energy E_{vac}, Fermi energy E_F and work function Φ as marked. The measured spectrum as a function of the kinetic energy E_{kin}^{out} of the ejected photoelectrons is shown on the top right side. b) Three-step model of the photoemission process reducing the photoemission event into three independent steps. (1) optical excitation between the initial E_i and final bulk Bloch eigenstates E_f, (2) travel of the excited electron to the surface, and (3) escape of the electron through the barrier potential of the surface into the vacuum. ((a) and (b) adopted from [103]).

sition from states of an occupied band into states of an unoccupied band. In the ARPES experiment ($h\nu = 5$ to $100\,eV$) the momentum of the photon is relatively small compared to the electron momentum, so that it can be neglected. Thus, the total momentum conservation which must apply in the photoemission process, requires that transitions in the reduced-zone scheme between the initial state with wave vector $\mathbf{k}_i^{in}$ and the final state with wave vector $\mathbf{k}_f^{in}$ are necessary k-conserving or vertical ($\mathbf{k}_i^{in} = \mathbf{k}_f^{in}$). In an extended-zone scheme, which is the more realistic description for a photoemission process, however, the transition from an initial state to a final excited state can also be connected over a reciprocal lattice vector G, i.e. $\mathbf{k}_f^{in} - \mathbf{k}_i^{in} = G$. Thus the crystal lattice provides the additional momentum an electron needs to reach the final state. This representation highlights the fact that a direct optical interband transition is a process involving diffraction against the lattice [39].

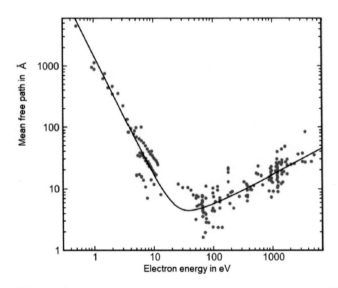

Figure 3.5: Measured mean free path of electrons as a function of energy (dots) and corresponding least square fit to the data (straight line). (Taken from [104]).

Resuming the three-step model, the transport of the excited electron to the surface is often accompanied with a loss of kinetic energy due to inelastic scattering processes. If the mean free path of the photoelectron (see Fig. 3.5) is lower than the initial distance from the surface, the electron may not escape the crystal. The dominant scattering event is hereby electron-electron interaction. This scattering process also leads to the appearance of secondary electrons, which only loose a part of their kinetic energy and may thus be visible at different kinetic energies in the measured band structure.

Finally, transmission through the surface of the photoelectron is achieved when the final state Bloch eigenstate matches to a free-electron plan wave propagating into the vacuum. Moreover, the kinetic energy of the electron normal to the surface, left after the transport to the surface, must overcome the work function barrier of the material in order to escape from the bulk. The total momentum of the photoelectron is not conserved as the electron crosses the surface ($\mathbf{k}^{in} \neq \mathbf{k}^{out}$). One reason is that the potential step at the surface reduces the component of the kinetic energy perpendicular to the surface. Another reason is that the periodic crystal potential is not existing in the vacuum so that the dispersion of the electron in the solid is different from that in the vacuum. Both facts apply on the perpendicular component of the wave vector $\mathbf{k}_\perp$, whereas the parallel component $\mathbf{k}_\parallel$ is conserved by passing through the surface as translation symmetry is not broken for this component. Hence, $k_\parallel^{in} = k_\parallel^{out}$, which can be written as (see Fig. 3.6) [101]:

$$k_\parallel^{in} = k_\parallel^{out} = \frac{1}{\hbar}\sqrt{2mE_{kin}^{out}}\sin\Theta. \tag{3.15}$$

with Θ being the angle of emission (polar angle) normal to the surface. As $k_\perp^{in}$ is not conserved but necessarily required for the determination of the electronic structure $E(k)$, a different approach is needed. One has to make some assumption regarding the dispersion of the final states in the crystal $E_f(k)$. In particular, a nearly-free electron description for the final bulk Bloch states is adopted [101], and it follows according to the energy scheme in Fig. 3.4 a):

$$E_f(k) = \frac{\hbar^2 k^{in^2}}{2m} - V_0 = \frac{\hbar^2(k_\parallel^{in^2} + k_\perp^{in^2})}{2m} - V_0. \tag{3.16}$$

If $E_f(k)$ is referenced to the vacuum level E_v, it is equal to E_{kin}^{out}. V_0 is the crystal potential as defined in Fig. 3.4 a) where E_0 denotes the bottom of the imaginary parabola defining the final states. With $\hbar^2 k_\parallel^{in^2}/2m = E_{kin}^{out}\sin^2\Theta$ from eq. 3.15 it thus follows for $k_\perp^{in}$:

$$k_\perp^{in} = \frac{1}{\hbar}\sqrt{2m(E_{kin}^{out}\cos^2\Theta + V_0)}. \tag{3.17}$$

If the crystal potential V_0 is known, the corresponding value for $k_\perp^{in}$ is thus achievable. However, as the determination of V_0 is often only given by theoretical calculations, the exact value of $k_\perp^{in}$ is mostly uncertain.

For particular cases however, the $k_\perp^{in}$ uncertainty is less relevant, as for example in low-dimensional systems which are characterized by an anisotropic band structure and negligible dispersion along the z-direction. The electronic structure is then predominantly determined by the parallel component of the wave vector. As a result, the simple tracking of the exit angle Θ of the photoelectron, together with its kinetic energy provides a detailed picture of the electronic dispersion $E(k)$ when applying eq. 3.14 and 3.15. The same is, of course, true for all surface states.

As already briefly discussed above, the photoemission results are often discussed within the three-step model (Fig. 3.4 b)), which has proven to be rather successful [105, 106, 102]. The total photoemission intensity is thus given by the product of three independent terms, i.e. the probability for the optical transition, the scattering probability of the traveling electron and the transmission probability through the surface barrier [101]. However, the major information of the intrinsic electronic band structure is mainly encoded in the

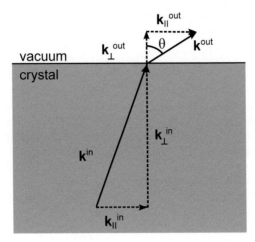

Figure 3.6: Evolution of the electron's momentum **k**, when passing trough the surface barrier. $\mathbf{k}^{in}$ and $\mathbf{k}^{out}$ describes the momentum of the electron inside and outside the bulk, with $\mathbf{k}_\parallel$ and $\mathbf{k}_\perp$ its parallel and perpendicular components, respectively. Θ denotes the polar angle at the exit of the crystal.

first step, where a transition probability $T_{i,f}$ for an optical excitation between an N-electron initial state $|u_i^N\rangle$ and a possible final state $|u_f^N\rangle$ can be deduced using Fermi's golden rule [101]:

$$T_{i,f} = \frac{2\pi}{\hbar} \left| \left\langle u_f^N \middle| H_{int} \middle| u_i^N \right\rangle \right|^2 \delta \left(E_f^N - E_i^N - h\nu \right), \qquad (3.18)$$

with E_i^N and E_f^N being the initial and final states energies of the N-particle system and H_{int}, the Hamiltonian which considers the interaction between the electron and the photon as a perturbation. The N-electron final state wave-function u_f^N is mostly fragmented into the wave-function of the photoelectron $\phi_f(\mathbf{k})$ and the final wave-function of the $(N\text{-}1)$-electron system u_f^{N-1}:

$$u_f^N = A\phi_f(\mathbf{k})u_f^{N-1}, \qquad (3.19)$$

with A, an operator which antisymmetrizes the N-electron wave-function in a way that the Pauli principle is satisfied. The same decomposition is valid for the N-electron initial state, such that the matrix elements from eq. 3.18 can be written:

$$\left\langle u_f^N \middle| H_{int} \middle| u_i^N \right\rangle = \left\langle \phi_f(\mathbf{k}) \middle| H_{int} \middle| \phi_i(\mathbf{k}) \right\rangle \left\langle u_f^{N-1} \middle| u_i^{N-1} \right\rangle, \qquad (3.20)$$

and $\langle \phi_f(\mathbf{k}) | H_{int} | \phi_i(\mathbf{k}) \rangle \equiv M_{i,f}(\mathbf{k})$ is defined as the one-electron dipole matrix element and the second term is the $(N\text{-}1)$-electron overlap integral. From this point, one can derive the total photoemission intensity as a function of a distinct $E_{\text{kin}}^{\text{out}}$ at a particular momentum k as the overall transition probability, i.e. $I(k, E_{\text{kin}}^{\text{out}}) = \sum_{i,f} T_{i,f}$, which is proportional to [101]:

$$I(k, E_{\text{kin}}^{\text{out}}) \propto \sum_{i,f} |M_{i,f}(\mathbf{k})|^2 \sum_f |c_{i,f}|^2 \delta \left(E_{\text{kin}}^{\text{out}} + E_f^{N-1} - E_i^N - h\nu \right). \quad (3.21)$$

$|c_{i,f}|^2 = \left| \left\langle u_f^{N-1} | u_i^{N-1} \right\rangle \right|^2$ here denotes the probability that the transfer of an photoelectron from an initial state i remains the $(N-1)$-electron system in the excited final state f. The photoemission intensity thus mainly depends on the one-electron matrix elements, for which special dipole selection rules apply, identifying the possible transitions between initial and final energy bands for each particular case. The selection rules deeply depend on the particular symmetry properties of bands in a crystal, so that contributions of specific bands in the photoemission spectrum can be a-priori been ruled out.

A more rigorous approach for the determination of the photoemission intensity is to consider the so-called *one-step model* where the whole photoemission process (step 1 to 3) is treated as a single coherent step [39]. In that way, contributions from bulk, surface and vacuum have to be considered in common in the Hamiltionian from eq. 3.18, which massively complicates the calculation of the photoemission event.

3.2.1 Spin-resolved ARPES

A very powerful property of the ARPES technique beside the acquisition of the electronic band structure is, that it can be used in order to measure the spin degree of freedom of particular bands in a solid. This mode is known as spin-resolved ARPES (spin-ARPES). The requirement for the experimental ARPES setup is the implementation of a Mott detector, which is able to detect the specific spin component of each photoelectron. The setup is depticted in Fig. 3.7. After the energy selective passing through the energy analyzer, the electrons are accelerated and directed towards a heavy metal foil (Au foil in the sketch) where they get scattered. Due to the strong SO coupling within the foil, the scattering of the photoelectrons depend on their spin and results in a separation of the distinct spin components. In Fig. 3.7 for example, the spin-up component is deflected to the left, and the spin-down component to the right. Special detectors which are placed in the different scattering directions count the selected photoelectrons. After calibration of the foil, the net spin-polarization in up-down direction can then be deduced from the difference in the photoelectron current of the left and right detector. The same can be done for the other spin directions. A typical data set from a spin-ARPES

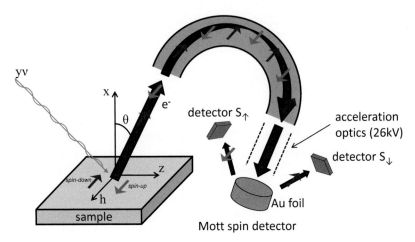

Figure 3.7: Sketch of the experimental setup for spin-resolved ARPES. Same geometry as for conventional ARPES (Fig. 3.3) but with a Mott detector placed behind the energy analyzer. The Mott detector consists of a heavy metal foil, which scatters the electrons depending on their spin orientation and thus leading to a separation of the different spin components. Corresponding detectors count the respective spin-polarized electrons whereas the difference in the photoelectron current between two opposite detectors then provides the net spin-polarization for one particular spin orientation.

measurement showing the resulting spin-polarization for a particular spin-direction has already been visualized in Fig. 2.11 d) of section 2.2.7.

However, the intensity in the spin-ARPES is much smaller than in conventional ARPES experiments, which is due to the low efficiency of the Mott detector. The counting rate is thereby reduced up to a factor of ten with respect to the counting rate in the ARPES experiment, necessarily leading to a much longer acquisition time in order to gain meaningful statistics. A further disadvantage of the spin-ARPES experiment is the insufficient instrumental resolution typically used to increase the spin-ARPES intensity, especially in measurements where different bands are relatively close to each other [75]. The measured spin-polarization may then be significantly reduced with respect to the real one. Especially in the case of topological surface states, where theory predicts a spin polarization of nearly unity [75], extrinsic factors in the spin-ARPES experiment complicates a detailed view on this assumption. However, by a rigorous analysis of the different bands and a reasonable subtraction of a non-polarized background from the spin-polarized data, a quantitative consideration of the *real* intrinsic spin-polarization is possible, at least, at certain photon energies, as will later been shown in this work.

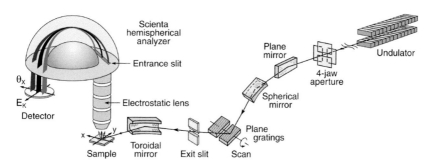

Figure 3.8: Typical beamline equipment showing the radiation path from the exit of the electron storage ring, where the radiation is produced, to the detector of the ARPES device. (Taken from [107]).

3.2.2 Experimental details: ARPES at the synchrotron

The ARPES and spin-ARPES experiments shown in this work have been performed at 300 K with electron analyzers Scienta R8000 and SPECS PHOIBOS 150 using linearly polarized synchrotron radiation from the beamlines UE112-PGM-1 and UE112-lowE-PGM2 at BESSY II in Berlin. A typical radiation path from the electron storage ring of the synchrotron, where the radiation is produced, to the detector in the ARPES chamber is visualized in Fig. 3.8. At the exit of the storage ring, a beam of white radiation is produced by an undulator and later momochromatized at the desired photon energy by a grating monochromator. The beam is focused on the sample and the emitted photoelectrons collected by the analyzer, where kinetic energy and emission angle are determined [107]. The whole system is kept at ultra high vacuum (UHV) at a base pressure of $1 \cdot 10^{-10}$ mbar. The big advantage of synchrotron radiation with respect to an ordinary gas-discharge lamp is, that the radiation covers a wide spectral range, from the visible to the X-ray region with a high intensity and switchable polarization. Typically, ARPES experiments are performed at photon energies in the ultraviolet regime ($h\nu < 100$ eV), in which the valence band structure of a material is accessible. In this energy range, the measurement is very sensitive to the surface as can be deduced from the energy dependence of the photoelectrons' mean free path (a few Å) (see Fig. 3.5). The core levels, however, are measured by X-ray photoemission spectroscopy (XPS), which operates at photon energies above 100 eV. Both information can be measured consecutively at the synchrotron only by changing the beam energy through the monochromator.

The analyzer used in the ARPES experiment is a Scienta hemispherical analyzer with, at the end, a two-dimensional position-sensitive detector, consisting of two microchannel plates and a phosphor plate in series, followed by a charge-coupled device (CCD) camera[2]. The advantage of this arrangement is, that it can be operated in angle-resolved mode and thus detecting the photoelectrons of different emission angles simultaneously. In particular, an angular window of photoelectrons, which is defined by the entrance slit and the electron optics in front of the entrance slit, is focused on different lateral positions on the detector (different colors correspond to different emission angles in Fig. 3.8). It is thus possible to map multiple energy distribution curves (EDCs) at the same time, generating a 2D picture of energy versus momentum. This is in strong contrast to a conventional electron analyzer, in which the momentum information is measured consecutively for adjacent k vectors and thereby each time averaged over all the electrons within the acceptance angle (typically $1°$). Hence, the acquisition time for a complete band structure $E(k)$ is much quicker and the angular resolution considerably higher.

The energy resolution in the ARPES experiment is basically given by the hemispherical deflector of the analyzer. The deflector consists of two concentric hemispheres of radius R_1 and R_2 with a potential difference of ΔV, so that only electrons within a narrow energy range, adjustable by a so-called pass energy E_{pass} will pass trough the hemispherical capacitor and thus reaching the CCD camera of the detector. The pass energy is given by:

$$E_{pass} = e\Delta V \left(\frac{R_1}{R_2} - \frac{R_2}{R_1} \right)^{-1}. \tag{3.22}$$

The energy resolution of the measured kinetic energy is then defined by [107]:

$$\Delta E = E_{pass} \left(\frac{w}{R_0} + \frac{\alpha^2}{4} \right), \tag{3.23}$$

with $R_0 = (R_1 + R_2)/2$, the width of the entrance slit w and the acceptance angle α. E_{pass}, w and α are all adjustable parameters which may be reduced in order to improve the energy resolution in the experiment. However, as one reduces these parameters, the intensity of the photon beam also decreases. Consequently, the best adjustment for the measurement is always a compromise between losing resolution and gaining beam intensity or vice-versa.

The momentum resolution of the experiment $\Delta k_\parallel$ can be derived from eq. 3.15 and is given by [107] (without argument):

$$\Delta k_\parallel \approx \frac{1}{\hbar} \sqrt{2m E_{kin}^{out}} \cos \Theta \cdot \alpha, \tag{3.24}$$

thereby neglecting the contribution of the finite energy resolution. Thus, the momentum resolution improves for lower photon energies (i.e. lower E_{kin}^{out}),

---

[2]For a detailed description see [107].

larger emission angles Θ and smaller acceptance angle α of the electron analyzer, again accompanied with a loss in beam intensity.

Typical energy and momentum resolution of a Scienta analyzer are a few meV and 0.2°, respectively. However, the energy and momentum resolution in the spin-ARPES experiment, which has been achieved with a Rice University Mott polarimeter operating at 26 kV in the conventional mode, are 100 meV and 1.4°, respectively.

For more information of synchrotron radiation technology and the development of the Scienta electrons spectrometers see references [108, 109, 110].

4 Identification of Tellurium based Phase-Change Materials as Strong Topological Insulators

Phase-change materials (PCMs) [111, 112] are a class of materials which have become of tremendous technological importance over the last two decades. Their ability of a fast and reversible phase transition between an amorphous and crystalline phase makes them appropriate for the application in data storage. PCMs based on Ge-Sb-Te (GST) alloys are characterized by a profound change of optical reflectivity and electrical conductivity upon changing from the amorphous to the crystalline phase [113, 114]. As a consequence, these alloys are already widely used in optical data storage, such as compact discs (CDs) or rewritable digital video discs (DVDs). The resistivity change upon crystallization, moreover, makes the GST alloys also a promising candidate for non-volatile electrical data storage, e.g. phase change random access memories (PCRAMs) [115, 116].

Thus, the optical and electronic contrast between the amorphous and crystalline phase as well as the fast, reversible switching on the nanosecond (ns) scale are the crucial requirements for materials to be a potential PCM for the technological application in data storage. A significant number of phase-change alloys built of different combinations of Ge-Sb-Te are mapped in the ternary phase diagram in Fig. 4.1 [117]. This map is constructed on the base of a fundamental understanding of bonding characteristics, as it was shown that resonance bonding in the crystalline state of a material is a unique fingerprint of PCMs [118]. Some of these PCM alloys, e.g. $Ge_1Sb_2Te_4$ (GST-124) and $Ge_2Sb_2Te_5$ (GST-225), lie on the so-called pseudobinary line between GeTe and Sb_2Te_3.

In the course of the discovery of the TIs, phase-change materials from the pseudobinary line have also been taken into consideration, as these materials face the main requirements for TI nature, namely built out of heavy elements, and thus having a strong SO interaction, and being a semiconductor with a band gap of several hundred meV. Sb_2Te_3 was one of the first predicted strong 3D TIs [67] (see section 2.2.7), and DFT soon after proposed a similar TI behavior for particular ternary compounds on the pseudobinary

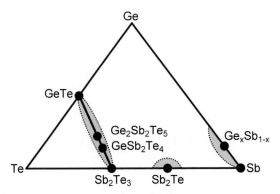

Figure 4.1: Ternary phase diagram of Ge, Sb and Te with PCMs marked by gray areas. Three groups, all having phase-change characteristics can be identified (gray areas surrounded by dotted lines). The most prominent group is the pseudobinary line between GeTe and Sb_2Te_3 with materials such as $Ge_1Sb_2Te_4$ and $Ge_2Sb_2Te_5$. (Taken from [117]).

line [119, 120, 121, 122, 123]. Within this chapter, an experimental view on this proposals will be provided, using surface sensitive techniques, able to map the topological fingerprints present in this particular phase change alloys. It starts with the already well-known material Sb_2Te_3 and follows up the pseudobinary line, looking at the most prominent ternary PCM, namely $Ge_2Sb_2Te_5$. Identifying TI properties in classes of materials already in use for electronic or storage applications is highly desirable and opens up new possibilities towards the utilization of these fundamental new properties.

4.1 Fundamentals of phase-change materials

As already mentioned above, the fast reversible switching between an insulating amorphous and a conducting crystalline phase is the main fingerprint of a PCM used in data storage application. Both phases are thermodynamically stable and can be switched repeatedly for a large number of cycles (more than 10^5 [111]). The mode of principle of data storage is shown in Fig. 4.2. A short laser or current pulse of high intensity locally (μm scale) anneals the PCM above its melting temperature T_m, followed by a rapid cooling of the area at $10^9\,\mathrm{Ks^{-1}}$. This brings the area into a disordered, amorphous phase and marks the *write*-event in the storage process. The amorphous phase has a different optical contrast than the surrounding crystalline state and is thus detectable by a low-intensity laser beam (*read*-event). Adjusting a long laser or current pulse of low intensity which locally anneals the PCM above its

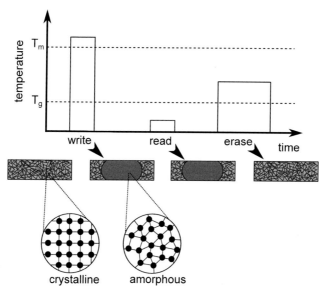

Figure 4.2: Scheme of the rewritable optical data storage process. From left to right: a short, high laser pulse heats the PCM over its melting temperature T_m. Subsequent rapid cooling at rates higher than $10^9 \, Ks^{-1}$ quenches the state into a disordered, amorphous phase. This step marks the *write*-event. A low-intensity laser beam detects the different optical contrast between the crystalline and amorphous phase and marks the *read*-event. The crystalline phase is restored by annealing the PCM over the crystallization temperature T_g and marks the *erase*-event of the storage process. (Adopted from [126]).

crystallization temperature, often called the glass temperature T_g, brings the state back into the crystalline phase and marks the *erase*-event of the storage process. The laser or electrical heat induced switching occurs within nanoseconds [114] or below [124] at a very low energy cost of only 1 fJ [125].

The first materials showing fast recrystallization and a good optical contrast were GeTe [127] and $Ge_{11}Te_{60}Sn_4Au_{25}$ [128, 129]. This led to the discovery of pseudobinary alloys along the $GeTe-Sb_2Te_3$ tie line (Fig. 4.1), i.e. GST-147, GST-124 and GST-225 [114]. The origin of the optical contrast between the amorphous and crystalline phase in GST alloys, however, is far from being settled [130]. Measurements of the dielectric function ϵ of GST-124 using infrared spectroscopy and spectroscopic ellipsometry, reveal an optical dielectric constant which is 70-200 % larger for the crystalline than for the amorphous phases, depending on the excitation energy which ranges from 0.025 to 3 eV in the experiment [118]. This difference is attributed to a significant change in bonding between the two phases. In the crystalline phase of GST alloys, resonance bonding occurs, which is due to a discrepancy between the number of electrons and the number of nearest neighbors engaged in a bonding. The result is a necessary superposition of two covalent bonding configurations with strongly delocalized electrons. As a consequence, a significantly increased polarizability is present, which does not take place in the amorphous phase, where resonance bonding is absent [118].

The atomic structure of the amorphous and crystalline phase, as well as the structural changes upon switching between the phases are yet not fully understood. However, a huge amount of experimental analysis in recent years have led to a better understanding, at least for the crystalline state in the GST compounds. Generally, the crystalline phase consist of two slightly different phases, i.e. a metastable cubic one, which is used for application [131], and a stable hexagonal one. A fingerprint of the metastable cubic structure in GST is the large amount of vacancies which is of the order of 10 % [111], and which is not found in other materials. Matsunaga *et al.* [132] found by X-ray diffraction (XRD) measurements that the metastable cubic phase of the GST compound consists of a rocksalt structure with two different sublattices. One sublattice only containing Te atoms and the other one built by statistically distributed Ge, Sb and vacancies. Matsunaga among others [133, 134] suggested that these vacancies, also present in the amorphous phase, play an important role in the fast switching between the two phases, namely by a movement of the vacancies during the structural transition.

Moreover, EXAFS (Extended X-ray Absorption Fine Structure) measurements [136] and DFT calculations [137] propose that the rocksalt structure in the metastable crystalline phase differs slightly from the ideal rocksalt structure. DFT reveals pronounced local distortions for the nearest-neighbor Ge-Te bonds, which lead to a splitting of those bonds in shorter and longer Ge-Te bonds. A similar finding is observed for the Sb-Te bonds, even though the splitting into shorter and longer bonds is less pronounced. Such a splitting into shorter and longer bonds is often denoted as a Peierls effect, a well-known phenomenon in many binary chalcogenides [133]. The distortions are smallest for GST-224 (10 pm in average) and largest for GST-124 (18 pm in average) [133]. The stable hexagonal phase of GST differs from the metastable one mainly by its vacancy distribution, which forms an own vacancy layer instead of being randomly distributed along the lattice. Matsunaga *et al.* [138] proposed a hexagonal lattice, build by a 9-layer block along the *c*-axis which are periodically spaced by van-der-Waals gaps (described in more detail in section 4.3.1 and Fig. 4.23 d)). Similar models are proposed by Petrov *et al.* (Petrov phase) [139] and Kooi *et al.* (KH phase) [140] but with Ge and Sb atoms forming separate layers. Both, the Petrov phase and the KH phase differ by their respective stacking of the pure layers. We will later see that the

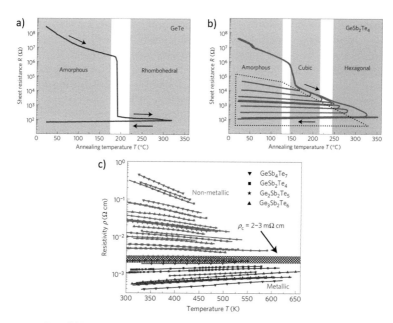

Figure 4.3: a) and b) Temperature dependence of the sheet resistance of 80 nm GeTe film and 100 nm GST-124 film measured in van der Pauw geometry, respectively. Arrows mark direction of cooling or heating. c) Resistivity for different GST alloys plotted during the cooling down for the region marked by dotted line in (b). At a critical resistivity (ρ = 2-3 mΩcm) the behavior changes from non-metallic (dρ/dT < 0) to metallic behavior (dρ/dT > 0). (Taken from [135]).

specific stacking order of the layers is crucial for the appearance of TI properties [120, 123] (section 4.3.1).

Until now, the driving mechanism leading to the ultra-fast switching between the amorphous and crystalline phase is far from being settled. Several models have been proposed, with most of them suggesting close structural similarities in both phases so that only small bond-rotational movements over a few relevant atoms are necessary to switch from the amorphous to the crystalline phase [136, 134, 141, 124]. This implies the presence of well structured building blocks, switching as a whole during the transition, with an overall medium-range ordering of the atoms already present in the amorphous phase. Welnic *et al.* [126] therefore proposed a spinel structure to be a suitable candidate from DFT, with the Ge atoms occupying tetrahedral positions, and the Te and Sb atoms occupying octahedral positions. An interesting proposal which is sustained by EXAFS data is given by Kolobov *et al.*, claiming that

Ge atom switches from the tetrahedral position (amorphous phase) into a octahedral position (crystalline phase) just by adding sufficient energy to the
system. Such a structural transformation involves a change in the hybridization from sp3-hybridization in the amorphous state to p-type bonding in the
rocksalt metastable crystalline state, also providing a possible explanation for
the strong electrical contrast [136]. Moreover, it has been observed that the
structural transition is accompanied by a significant reduction of the density
[142, 143, 144]. The volume of GST-124 has for example been documented to
decrease by about 5% upon crystallization [144].

Besides the strong optical contrast being present in PCMs, there is also a
significant electrical contrast present in GST alloys which may be used for
non-volatile electronic data storage (e.g. PCRAM) [115]. In many PCMs, depending on the particular stoichiometry, the conductivity increases by more
than three orders of magnitude upon crystallization, ensuring a high signal-
to-noise ratio [113, 145]. Resistivity measurements on as-deposited thin films
of GeTe and GST-124 upon annealing from the amorphous to the crystalline
phase show different behavior, as depicted in Fig. 4.3 a) and b) [135]. For
GeTe, which is at the border of the pseudobinary line, the resistivity sharply
drops during the transition without showing further temperature dependence in the crystalline phase. In contrast, the GST-124 film reveals a pronounced annealing dependence, with a drop in resistance by a factor of 400
in the temperature range between 150 to 350 °C. Interestingly, the GST-124
alloy here changes from non-metallic to metallic after the achievement of a
critical resistance, as can be observed by the different slopes of the resistivity
($dR/dT < 0$ or $dR/dT > 0$, respectively) upon cooling down (region marked
by dotted lines in Fig. 4.3 b)). The same behavior is also observed for other
GST alloys (Fig. 4.3 c)) and a critical resistivity of $\rho = 2\text{-}3 \, m\Omega cm$ for the metal-
insulator transition (MIT) is extracted [135]. Siegrist et al. [135] suggest the
high degree of disorder [146] in the crystalline phase of GST alloys responsible for the insulating behavior at low annealing temperature, despite the
high carrier concentration and the p-type conductivity, so that they behave
like an Anderson insulator [43]. GeTe, in contrast, lacks a comparable degree
of disorder, hence always being metallic. The transition from insulating to
conducting behavior in GST is finally explained by an increase of the charge-
carrier mobility upon annealing at almost constant charge-carrier density. A
similar result on as-deposited GST-124 is found by Subramaniam et al. [144].
In this work, STS measurements reveal a continuous decrease of the band
gap width with increasing temperature, and a closing of the gap in the stable
hexagonal phase. Moreover, the Fermi level E_F is found to shift continuously
from midgap in the amorphous phase to the valence band edge in the stable
hexagonal phase, also indicating a transition from an insulating to a conductive state upon annealing. However in the STS experiment, possible band

bending effects at the surface which could lead to a shifting of E_F might play a role. The resistivity measurement, in contrast, are rather bulk sensitive.

4.2 Identification of topological insulator properties in crystalline Sb_2Te_3

The heavy element alloys Bi_2Se_3, Bi_2Te_3 and Sb_2Te_3 have been the first 3D TIs proposed with the simplest electronic structure, namely with one single spin polarized Dirac cone at the Γ-point [67] (see section 2.2.7). While the TI properties of Bi_2Se_3 and Bi_2Te_3 are already well established, the phase change material Sb_2Te_3 was rarely probed. First ARPES measurements on single-crystal Sb_2Te_3 suggested that the Fermi level is within the bulk valence band probably due to hole doping [72]. No sign of a surface Dirac cone has been observed, however, the TI nature of the system has been indirectly deduced from the presence of an *inverted* bulk valence band, namely a band with a minimum at the Γ-point. This behavior is a hint for an inversion of bulk bands around E_F induced by strong SO coupling which thus often leads to the required parity change at the TRIMs. Stronger evidence for the TI nature of Sb_2Te_3 has been found by ARPES measurements on thin films grown by molecular beam epitaxy (MBE) [147]. The data reveal the lower part of a Dirac cone located around the Γ-point, and shows that the Dirac point is accessible by doping the thin films with cesium. This leads to an electron doping of the surface and a subsequent reduction of the p-type nature initially present due to intrinsic defects. However, the spin chirality and the topological nature of the Dirac cone has not been tackled.

4.2.1 STM and STS characterization of Sb_2Te_3

In this chapter, STM measurements on single crystal Sb_2Te_3 samples are shown, which have been cleaved in UHV at a base pressure of $1 \cdot 10^{-10}$ mbar prior to the measurements, in order to obtain a clean and adsorbate-free surface. The cleavage process includes a Cu-tape which is pressed on the surface of the sample and which is holding a small wire forming a bow. The Cu-tape can later be removed inside the UHV chamber only by pulling the bow by a standard wobblestick leaving a freshly cleaved surface. The single crystal of Sb_2Te_3 mounted to the sample holder is shown in Fig. 4.4 a), together with the Cu-tape after a cleavage process. The STM measurements are performed using an etched tungsten tip (Fig. 4.4 b)) inside a UHV insert within a helium cryostat ensuring a sample temperature of 6 K and operating at B-field strengths up to 7 T in z-direction and 3 T and 0.5 T in x- and y-direction, re-

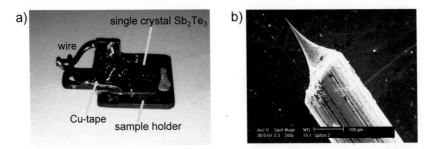

Figure 4.4: a) Image of the single crystal Sb_2Te_3 mounted to a tungsten sample holder. Cu-tape holding a wire has been pressed on the sample surface and is removed in UHV in order to cleave the sample prior to the STM measurements. The image shows the sample after cleavage. b) Electron microscope picture of an etched tungsten tip which is used for the STM measurements. ((b) taken from [148]).

spectively. A detailed description of the home-built UHV STM chamber and its specific components can be found in the references [148] and [149]. More information of the experimental details are found in section 3.1.3 and some of the results of this section are published in ref. [31].

The crystal structure of Sb_2Te_3 consists of consecutive quintuple layers (QLs), whereas one QL is built by the stacking sequence Te(1)-Sb-Te(2)-Sb-Te(1). The different numbers in parentheses mark the different environments of the Te layers. The atomic configuration is visualized in the sketch of the

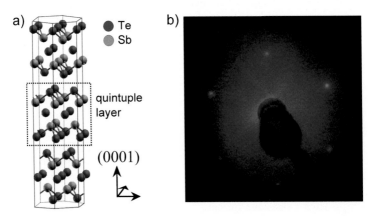

Figure 4.5: a) Sketch of the crystal structure of Sb_2Te_3; one QL is marked with different atoms in different colors as indicated. b) LEED pattern of the freshly cleaved Sb_2Te_3 (0001) crystal verifying the hexagonal symmetry of the Te terminated surface.

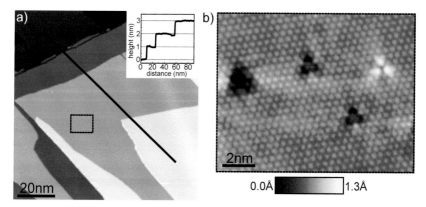

Figure 4.6: a) STM image of cleaved Sb$_2$Te$_3$ (0001) ($V = 0.9$ V, $I = 50$ pA) revealing large terraces with width of several 100 nm. Inset: Line profile showing the step heights which correspond to the height of one QL (≈ 1 nm). b) Atomically resolved STM image ($V = 0.4$ V, $I = 1$ nA) recorded in the area marked by the dashed box in (a). A hexagonally arranged pattern of the Te atoms is observed with an average atomic distance of 0.42 nm, which is in agreement with the theoretic value [150]. Types of defects are visible as triangular structures appearing dark and bright.

crystal structure in Fig. 4.5 a). The coupling within a QL is strong, whereas the interaction between two QLs is predominantly of van der Waals type [67]. Consequently, cleavage leads to a Te terminated (0001) surface with hexagonal symmetry as has been verified by low-energy electron diffraction (LEED)[1] (see Fig. 4.5 b)). Identically to Bi$_2$Se$_3$ and Bi$_2$Te$_3$, Sb$_2$Te$_3$ has inversion symmetry with the layer Te(2) containing the center of inversion. This simplifies the calculation of the $\mathbb{Z}_2$ topological invariant considerably, which becomes an analysis of states at the high symmetry points only, and, thus, leads to the straightforward identification of a strong topological insulator [13, 67] (see also section 2.2.5).

An STM overview image of the cleaved Sb$_2$Te$_3$ (0001) surface showing the topography of the crystal is depicted in Fig. 4.6 a). It reveals large terraces with width of several 100 nm separated by step edges of 1 nm in height, corresponding to the theoretical height of one QL. Further, the atomic arrangement of the surface is identified by zooming into a flat area on the terrace (dashed area in Fig. 4.6 a)). The corresponding STM image is shown in Fig. 4.6 b) and reveals an hexagonally arranged pattern of the Te-atoms with an atomic dis-

---

[1]LEED enables the determination of the symmetry of the surface structure of single-crystalline materials by bombardment with a beam of low energy electrons (20–200 eV). The diffraction pattern of the electrons is visible as spots on a fluorescent screen.

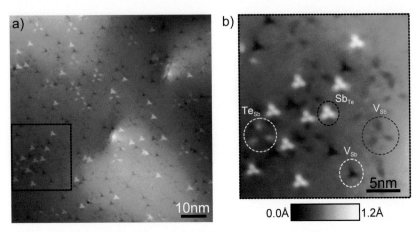

Figure 4.7: a) STM topographic image ($V = 0.9$ V, $I = 100$ pA) revealing typical types of defects for Sb_2Te_3. b) Zoom into the marked area of (a) showing a close up view of the four different types of defects (marked by different colored ellipses).

tance of $a = 0.42$ nm, nicely agreeing with the theoretical value [150]. Even in this small scale image, a significant amount of intrinsic defects states on the surface can be observed, which has already been identified to be responsible for the natural p-type conductivity of Sb_2Te_3 [73]. Especially defects, which act as acceptors lead to the p-type nature of the system. First-principles calculations in the work of Jiang *et al.* [73] identified the Sb vacancies (V_{Sb}) and the Sb-on-Te antisites (Sb_{Te}) to be the primary source of the p-type nature, whereas Te-on-Sb antisites (Te_{Sb}) act as a donor and lead to a natural n-type conductivity. Figure 4.7 a) shows an overview image of the single crystal Sb_2Te_3 surface, resolving four different types of defects. These defects are exposed in the close-up view of Fig. 4.7 b) and are identified based on the characterization by Jiang *et al.* [73]. Two types of Sb vacancies are present (V_{Sb}), showing each a depression at positive sample voltage and sitting on different atomic layers of the QL. The two antisite defects Sb_{Te} and Te_{Sb} show a bright contrast and are marked accordingly. It gets obvious from the STM data that the electron donor defects Te_{Sb} are significantly less present than the acceptor defects, explaining the rather natural p-type nature of Sb_2Te_3. The same tendency is found on other areas of the sample, always outnumbering the acceptor defects. Obviously, the exact strength of the p-type conductivity critically depends on the respective number of specific defects and may vary on the local scale.

Before we have a look at the STS results which provide information about the p-type nature of the sample, as well as possible hints for TI properties,

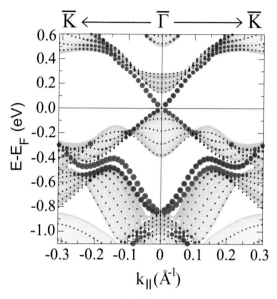

Figure 4.8: Band structure of Sb_2Te_3 in $\overline{\Gamma} - \overline{K}$ direction as calculated by DFT including spin-orbit coupling; states resulting from a film calculation are shown as circles with the color (blue or red) indicating different spin directions and the size of colored circles marking the magnitude of the spin density (for absolute spin polarization values see Fig. 4.14); shaded areas are projected bulk bands originating from a bulk calculation. (Calculation by Gustav Bihlmayer).

the electronic band structure of Sb_2Te_3 as deduced from DFT calculation will be introduced. The calculations are performed by Gustav Bihlmayer within the generalized gradient approximation [151] to DFT, employing the full-potential linearized augmented plane-wave method as implemented in the Fleur code[2]. SO coupling is included in a non-perturbative manner [152]. Based on the optimized bulk lattice parameters, the surfaces are simulated by films of a thickness of six QLs embedded in vacuum.

The band structure from DFT calculations including SO interaction is shown in Fig. 4.8. It combines a surface calculation, which is deduced from a six QL thick structure, and a bulk calculation, which is displayed as pro-jected bulk bands painted as gray shaded areas. The overall calculation is in good agreement with a recent calculation [153]. Small differences to ear-lier calculations [67, 72] can be traced back to the sensitivity of the electronic structure to small changes of the geometrical parameters. The spin-polarized

---

[2]for a program description, see http://www.flapw.de

surface states resulting from the surface calculation are displayed as colored circles being blue or red for the different spin orientations. Only the spin polarization perpendicular to the in-plane wave vector of the electrons $k_\parallel$ and the surface normal is shown. The varying radius marks the absolute value of the k-resolved spin density at, and above the surface. The absolute spin polarization of the states in comparison with the values from the experiment is discussed in more detail in Fig. 4.14 of the section 4.2.2. A single Dirac cone originating from topological protection is visible around $\overline{\Gamma}$ with Dirac point at E_F. The lack of a p-type conductivity results from the fact that defects are not present in the calculation. Strikingly, an overlap of the occupied Dirac surface states with the bulk states is observed, whereas the upper part of the Dirac cone resides in the bulk band gap. Furthermore, the $\mathbb{Z}_2$ topological invariant was checked by DFT and found to be topologically non-trivial with $\nu_0;(\nu_1\nu_2\nu_3) = 1;(000)$. Interestingly, another bulk "band gap" region exists for the projected bulk band in the calculation around $E - E_F = -400\,\text{meV}$ and at $k_\parallel$-values between $-0.3\,\text{Å}^{-1}$ and $+0.3\,\text{Å}^{-1}$. It houses two spin-polarized surface states exhibiting a Rashba-type spin splitting $\Delta E = \alpha \cdot |k_\parallel|$ close to $\overline{\Gamma}$ with the wave number parallel to the surface $k_\parallel$. Further, the Rashba coefficient has been deduced from the calculation to be $\alpha \simeq 1.4\,\text{eVÅ}$, at least, up to $k_\parallel \simeq 0.05\,\text{Å}^{-1}$. This α is larger than the value for Au(111) ($\alpha = 0.33\,\text{eVÅ}$) [22] or Bi(111) ($\alpha = 0.55\,\text{eVÅ}$) [26], both consisting of heavier atoms, but lower than the largest α-values so far found in Bi surface alloys ($\alpha = 3.8\,\text{eVÅ}$) [25].

Next, the electronic structure of Sb_2Te_3 is probed by STS as shown in the inset of Fig. 4.9 a). As already described in section 3.1.3, STS records the differential tunneling conductivity dI/dV which is proportional to the local density of states (LDOS) of the sample [100]. A decrease of the dI/dV signal at around 75 meV accompanied with a significant increase at around 230 meV is observed in the spectrum. This energy range approximately marks the band gap area and fits in size with the band gap found in the DFT (Fig. 4.8) and in the theoretical predicted gap by Zhang et al. [67]. Thereby, a minimum in the conductance which is approximately located around 170 meV above E_F is identified and attributed to the energy position of the Dirac point (E_D). Further, a considerable p-type nature of the single-crystal sample is observed, which arises from the large amount of intrinsic defects as discussed above, and which agrees with a recent STM study on MBE grown thin films [84].

The Dirac fermion nature of the topological surface states as shown in the DFT calculation (Fig. 4.8) can be experimentally demonstrated by the appearance of Landau levels in the $dI/dV(V)$ spectrum in the vicinity of an applied magnetic field B_z, pointing perpendicular to the sample surface. The energy of the electrons are thereby quantized into discrete values E_n. In contrast to the Landau levels of conventional electrons, a hallmark of Dirac fermions, is

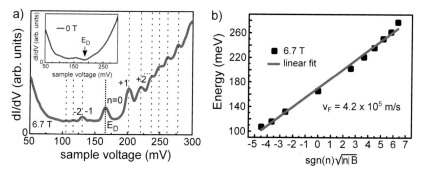

Figure 4.9: a) $dI/dV(V)$ spectrum ($V_{stab} = 0.3$ V, $I_{stab} = 400$ pA, $V_{mod} = 4$ mV) showing Landau quantization of the topological surface states at 6.7 T. E_D marks the position of the Dirac point located at the $n = 0$ Landau level. Inset: $dI/dV(V)$ spectrum ($V_{stab} = 0.3$ V, $I_{stab} = 50$ pA, $V_{mod} = 4$ mV) of Sb$_2$Te$_3$ without magnetic field. b) Landau level energies at $B_z = 6.7$ T plotted against $\text{sgn}(n)\sqrt{|n|B}$. The line is a linear fit to the data and the resulting Fermi velocity v_F is marked.

that a field-independent Landau level appears at the Dirac point. Furthermore, the energy position of the nth Landau level E_n of Dirac fermions has a square-root dependence with respect to the magnetic field and is expressed by [47, 154, 79, 83]:

$$E_n = E_D + \text{sgn}(n)\sqrt{2e\hbar v_F^2 |n|B}, \quad \text{with} \quad n = 0, \pm 1, \pm 2, \dots \quad (4.1)$$

where v_F is the Fermi velocity, e the electron charge and n the index of the specific Landau level. Figure 4.9 a) shows the $dI/dV(V)$ spectrum measured at a magnetic field of $B_z = 6.7$ T. A quantization of energy states into Landau levels is clearly visible, accompanied with an nonequally spacing of the Landau level peaks. A major peak is found at a sample voltage of about $V = 166$ mV, which is identified to be the field-independent $n = 0$ Landau level and thus the energy position of the Dirac point E_D. This also agrees with the Dirac point assignment in the $dI/dV(V)$ spectrum at 0 T. Below the zero mode, three further peaks can be identified, however with a nearly vanishing intensity. This might probably be due to the coupling of the surface states of the lower Dirac cone with the bulk valence band (BVB) as these states nearby overlap (see Fig. 4.8). This problem will be further discussed in section 4.2.2. On the other side, the Landau levels with a positive index n are much more pronounced as the corresponding surface states are located within the bulk energy gap, so that the scattering probability is reduced significantly, and thus leading to a larger lifetime.

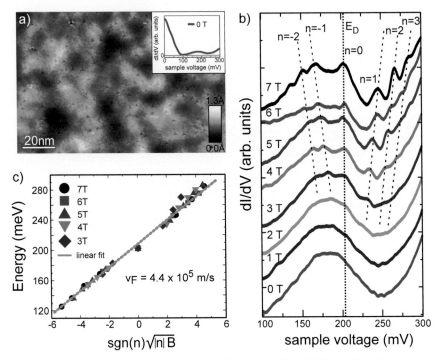

Figure 4.10: a) Large scale STM topography image ($V = 0.3$ V, $I = 50$ pA) showing corrugations and surface defects. Inset: $dI/dV(V)$ spectrum ($V_{stab} = 0.3$ V, $I_{stab} = 100$ pA, $V_{mod} = 2$ mV) taken away from the defects and without magnetic field. b) A series of $dI/dV(V)$ spectra ($V_{stab} = 0.3$ V, $I_{stab} = 100$ pA, $V_{mod} = 2$ mV) from $B_z = 0$ to 7 T of a narrow energy range taken on the same area as the spectrum in a). The vertical dotted line indicates the field-independent $n = 0$ Landau level at the Dirac point. The spectra are shifted vertically for clarity. c) Landau level energies for magnetic fields from 3 to 7 T plotted against $sgn(n)\sqrt{|n|B}$. The dotted line is a linear fit to the data and the resulting Fermi velocity v_F is marked.

The Dirac fermion nature of the detected electron states can be proven by plotting the energies of the peak positions E_n versus $sgn(n)\sqrt{|n|B}$ (eq. 4.1). For this purpose, the peak positions of the Landau levels have been determined by fitting the different peaks by Lorentzian functions. The result is plotted as single squares in Fig. 4.9 b) together with a linear fit to the extracted data. For the first three Landau levels on each side of the $n = 0$ Landau level, the straight line fits quite good, confirming the Dirac fermion type nature of the surface state near the Dirac point. However, towards higher Landau level indexes, the dispersion slightly deviates from linearity and moves into a convex function as also observed in the DFT calculation. The resulting Fermi velocity is extracted to $v_F = 4.21 \pm 0.13 \cdot 10^5$ m/s, which agrees reasonably with $v_F = 3.2 \cdot 10^5$ m/s obtained by DFT (Fig. 4.8).

A further series of Landau level spectroscopy has been performed on a different location of the sample (STM image in Fig. 4.10 a)) with a similar defect density, however mostly of the V_{Sb} type, whereas other types of defects are hardly visible. The corresponding $dI/dV(V)$ spectrum which was measured for a larger energy range and without magnetic field is displayed in the inset, showing two local minima, namely at energies of ≈ 100 meV and ≈ 225 meV. The $dI/dV(V)$ spectra revealing the Landau levels for different magnetic strengths are shown in Fig. 4.10 b). At a magnetic field of B_z = 3 T, small quantization peaks get visible, which become more pronounced with increasing field. A field-independent peak is found to be located at E = 208 meV (vertical dotted line in Fig. 4.10 b)) again identified as the Dirac point energy E_D and implying an even higher p-type doping as in the measurement before. Interestingly, the Dirac point is not found at the minimum of the $dI/dV(V)$ curve assuming that there are further contributions originating from the specific probing micro-tip in this STS experiment. Another explanation could be a possible overlapping of the Dirac point with the BVB so that the Dirac point is not fully detached. In this case, the bulk would contribute to an enhanced intensity at the Dirac point. However, from the DFT calculation, the BVB is overlapping with the lower part of the Dirac cone but not with the Dirac point. Moreover, as the energy resolution of the experiment is of the order of several meV (eq. 3.11), a bad resolution being responsible for a smearing of the Dirac point and the BVB can at least be excluded.

The Landau level energies E_n for the different magnetic fields (Fig. 4.10 b)) are again plotted versus $\mathrm{sgn}(n)\sqrt{|n|B}$ and linearly fitted. The resulting plot is displayed in Fig. 4.10 c). The Dirac fermion nature of the surface state electrons is reconfirmed and a Fermi velocity of $v_F = 4.44 \pm 0.07 \cdot 10^5$ m/s deduced, confirming the former value. Thus, the STS measurements on the single crystal Sb$_2$Te$_3$ demonstrate the Dirac fermion nature of the topological surface states by revealing the linear dispersion of the surface states as well as the field-independent n = 0 Landau level of the Dirac point.

An important observation of the two Landau level measurements shown above is, that the different energy position of the Dirac point differs when measured on a different area of the sample. Between the two measurements, E_D differs about 40 meV, pointing to local potential fluctuations within the sample surface, due to different types and amount of defects which vary on a local scale throughout the sample surface. Moreover, the electrostatic induction by the electric field which reigns between the STM tip and the sample also considerably shifts the bands on the surface as has been observed in 2D systems with low electron density [155]. Furthermore, it has been reported in a previous work on the TI Bi_2Se_3 [79], that the effect critically depends on the applied voltage between tip and sample which can lead to discrepancies of 100 to 200 meV between Dirac point energies, measured either by STM or ARPES.

4.2.2 Spin-resolved ARPES measurements of Sb_2Te_3

Characterization of the topological surface states in the fundamental gap

In the section above, I have already provided strong evidence for the presence of surface states which behave like 2D massless Dirac fermions on the surface of single crystal Sb_2Te_3. The major fingerprint of a TI however, namely the spin-polarized nature of the surface states has not been tackled so far. It is its special spin structure that characterizes a topological surface state (TSS). In this section, spin-resolved photoemission spectroscopy (spin-ARPES) is used in order to identify the spin nature of the states and thus provide final experimental evidence for the presence of TI properties in Sb_2Te_3. The measurements have been carried out at the synchrotron BESSY in Berlin in collaboration with the group of Prof. Dr. Oliver Rader and the experimental details

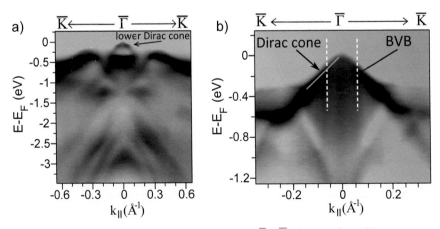

Figure 4.11: a) ARPES data of Sb_2Te_3 (0001) along $\overline{\Gamma} - \overline{K}$ at an incident photon energy $h\nu = 50$ eV; Dirac cone is marked. b) ARPES data at $h\nu = 55$ eV showing a close-up view of the lower Dirac cone along $\overline{\Gamma} - \overline{K}$ and the bulk valence band (BVB) as marked. Straight line is a guide to the eye from which the Fermi velocity v_F is deduced; dashed lines mark the positions where the spin-resolved energy distribution curves (EDCs) are measured (see Fig. 4.12 (a), (b) and (c)).

have already been described in section 3.2.2. The results are largely published in ref. [29].

Figure 4.11 a) shows a large energy scale ARPES image of Sb_2Te_3 after cleavage in UHV at a photon energy of $hv = 50\,eV$ and along the $\bar{\Gamma} - \bar{K}$ direction. The cut along a specific direction in **k**-space has been adjusted by LEED prior to the ARPES measurements. Distinct bands are visible in the data, especially two bands forming a cone-like feature close to E_F. If one zooms into this feature by mapping a closer energy range below E_F, two linearly dispersive bands crossing at $\bar{\Gamma}$ are found and which form a Dirac point exactly at E_F. This feature is most easily visible at $hv = 55\,eV$, revealing a relatively strong dependance of the bands on the specific photon energy. The observation of the Dirac point close to E_F indicates that in this case, the position of the surface Fermi level is predominantly determined by the Dirac electrons and not by intrinsic doping. This result deviate from our STM measurements of about 200 meV but can be explained by the strong local potential fluctuation on the surface which alone may already lead to local differences of several tens of meV [82]. Another reason for the discrepancy has already been discussed in the previous section and might be due to the STM tip induced band bending present for 2D systems with low electron density. However, our data also deviates from previous ARPES results obtained on bulk Sb_2Te_3 [72], but is in agreement with ARPES data from thin films grown by MBE [55]. This points to a low defect density of the investigated crystal with respect to other single crystal Sb_2Te_3 samples [73].

The linear dispersion of the Dirac cone in Fig. 4.11 b) is fitted by $E - E_F = \hbar v_F |k_{\parallel}|$ resulting in a Fermi velocity of $v_F = 3.8 \pm 0.2 \cdot 10^5\,m/s$ (straight line). This agrees reasonably with $v_F = 4.44 \pm 0.07 \cdot 10^5\,m/s$ from the STM measurements (Fig. 4.10 c)), as well as with $v_F = 3.2 \cdot 10^5\,m/s$ obtained by DFT (Fig. 4.8). Moreover, the background of the bulk valence bands which has been found in DFT is also nicely recovered in the ARPES data and reveals a possible overlap with the bottom part of the Dirac cone.

In the following, we will have a closer look at the spin nature of the TSS in order to verify the chiral spin polarization of the Dirac cone. Therefore, spin-ARPES is applied, which, in the set-up that was used at the synchrotron, is capable of measuring the spin component within the surface plane by recording **k**-specific energy distribution curves (EDCs).

Figure 4.12 a), b) and c) show the spin-resolved EDCs, measured at opposite positions of the lower Dirac cone, namely at $k_{\parallel} = -0.06\,\text{Å}^{-1}$ and $k_{\parallel} = 0.06\,\text{Å}^{-1}$ (as marked with dashed lines in Fig. 4.11 b)). The spin component perpendicular to $k_{\parallel}$ and the surface normal (Fig. 4.12 a) and b)) exhibits an intensity difference between spin-up and spin-down component which reverses for the opposite momentum. In contrast, the spin component parallel

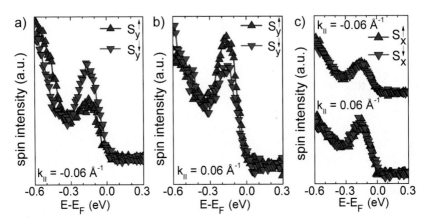

Figure 4.12: a), b) Spin-resolved energy distribution curves (EDCs) at $hv = 54.5\,\text{eV}$ for the spin component perpendicular to $\mathbf{k}_\parallel$ recorded at $k_\parallel$-values as indicated and marked by dashed lines in Fig. 4.11 b). The different colors mark different spin directions. c) Spin-resolved EDCs at $hv = 54.5\,\text{eV}$ for the spin component parallel to $\mathbf{k}_\parallel$.

to $\mathbf{k}_\parallel$ (S_x, Fig. 4.12 c)) shows no spin polarization. This leads to the spin momentum relation depicted in Fig. 4.13 b) for the in-plane spin component. The spin is perpendicular to $\mathbf{k}_\parallel$ and rotates counterclockwise for the lower part of the Dirac cone as also measured for Bi_2Te_3, Bi_2Se_3 [68, 71, 69, 72] or the tunable topological insulator $BiTl(S_{1-\delta}Se_\delta)_2$ [156]. The same sense of rotation is also found by the DFT calculation (Fig. 4.8).

The resulting spin polarization for the spin component S_y perpendicular to $\mathbf{k}_\parallel$ can be calculated according to:

$$P_y = (S_y^\uparrow - S_y^\downarrow)/(S_y^\uparrow + S_y^\downarrow), \qquad (4.2)$$

with $S_y^\uparrow$ and $S_y^\downarrow$ being the spin-resolved intensities perpendicular to $\mathbf{k}_\parallel$ from Fig. 4.12 a) and b). The result for the two opposite momenta $k_\parallel = -0.06\,\text{Å}^{-1}$ and $k_\parallel = 0.06\,\text{Å}^{-1}$ is displayed in Fig. 4.13 a). The net polarization is found to be $P_y \simeq 20\,\%$ with opposite sign for opposite momenta. For surface states well separated from the bulk states, it has been demonstrated that the spin polarization reaches unity [75]. However, as the BVB in Sb_2Te_3 overlaps with the lower part of the Dirac cone, a reduction of the spin polarization is likely. In order to analyze this, we have a closer look at the net spin polarization of the Dirac cone as predicted by DFT.

In this sense, the spin polarization of the Dirac cone in Sb_2Te_3 is calculated by DFT and analyzed in terms of an in-plane and out-of-plane component.

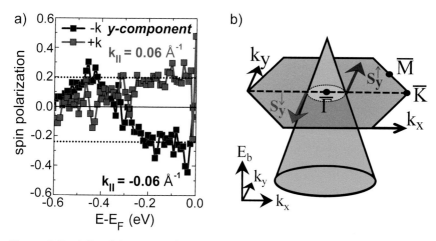

Figure 4.13: a) Resulting spin polarization perpendicular to the two different $k_{\parallel}$ as marked (EDCs from (a) and (b) in Fig. 4.12). b) Sketch of the lower Dirac cone with the spin directions marked as deduced from spin-ARPES and in accordance with DFT.

Only the in-plane component perpendicular to $\mathbf{k}_{\parallel}$ is considered, since no spin polarization was found for the direction parallel to $\mathbf{k}_{\parallel}$. Figure 4.14 shows the resulting spin polarization values with respect to the wavenumber integrated over the first two atomic layers. This area approximately corresponds to the penetration depth in the ARPES experiment at a photon energy of 55 eV (cf. Fig. 3.5). While $P_y \simeq 1$ is found for free Dirac cones by DFT [157], a reduced polarization for the lower Dirac cone of roughly 80 % near the $\overline{\Gamma}$ point is observed, which increases to about 90 % at $k_{\parallel} = 0.06$ Å^{-1}. This is mostly due to a penetration of the Dirac cone states into subsurface layers (Fig. 4.16) where fluctuating electric fields lead to a complex spin texture. In contrast, the in-plane polarization of the upper Dirac cone decreases towards higher wavenumbers down to 60 %. Thus, the vicinity of the bulk bands seems not to have an affect on the spin polarization of the lower Dirac cone, pointing to the fact that both, the TSS and the BVB are rather decoupled. Note the considerable out-of-plane polarization in $\overline{\Gamma} - \overline{K}$ direction for higher wavenumbers, which is in line with the warping of the Dirac cone at higher energies [71, 158]. This result agrees also qualitatively with the calculations of Yazyev *et al.* [159] for Bi₂Te₃ and Bi₂Se₃. However, the polarization values in those materials are typically smaller, reflecting the stronger spin-orbit entanglement caused by the heavier Bi atom.

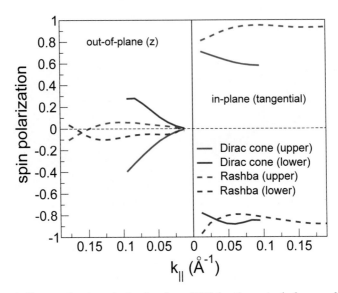

Figure 4.14: Expected spin polarization from DFT for the out-of-plane and in-plane component of the Dirac cone and the Rashba-type surface state (Rashba) along $\overline{\Gamma} - \overline{K}$. Upper and lower indicates the energies above and below the Dirac point and the energetically higher and lower Rashba band, respectively. The in-plane component considers exclusively the spin direction perpendicular to $k_\parallel$. (Calculation by Gustav Bihlmayer).

The discrepancy in the in-plane spin polarization between calculation ($P_y \simeq$ 90 %) and experiment ($P_y \simeq 20\,\%$) at $k_\parallel = 0.06\,\text{Å}^{-1}$ can thus be traced back to the finite angular resolution of the spin-ARPES experiment leading to a reduced spin polarization of the TSS due to the contribution of the unpolarized background of the BVB. Later, I will show that deconvolution from the BVB reveals an estimated spin polarization for the lower Dirac cone which is in a good agreement with the DFT result.

Characterization of the Rashba type surface state

The DFT calculation shown in Fig. 4.8 revealed a second spin-polarized state originating from SO interactions located in a gap at higher binding energies, which is of a Rashba type surface state (Rashba SS). It behaves similarly to the TSS, e.g. one spin branch connects the upper bulk band with the lower bulk band from -$k_\parallel$ to +$k_\parallel$ whereas the opposite spin branch acts in the opposite way. Figure 4.15 a) shows ARPES data along $\overline{\Gamma} - \overline{K}$ recorded at lower photon

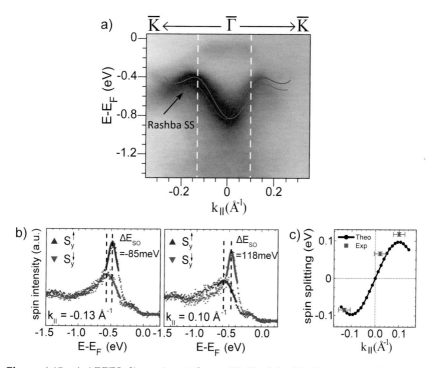

Figure 4.15: a) ARPES dispersion at $hv = 22\,\text{eV}$ of the Rashba type surface state (Rashba SS) along $\overline{\Gamma} - \overline{K}$ direction. The band structure of the Rashba surface states from DFT is superimposed as blue and red lines. Dashed white lines mark the position of the EDCs in (b). b) Spin-resolved EDCs (points in red and blue for the two spin directions perpendicular to $\mathbf{k}_{\parallel}$) at different momenta as indicated and marked by dashed white lines in (a). The Lorentzian fits of the peaks are shown as solid lines (red, blue) and the peak positions are marked by dashed lines. The spin splitting energies between two bands ΔE_{SO} is indicated. c) Calculated spin splitting (Theo) of the Rashba state in comparison with measured spin splitting (Exp). (Calculation in (c) by Gustav Bihlmayer).

energy, i.e. $hv = 22\,\text{eV}$. A prominent band is visible between -0.4 and -0.8 eV and exhibits an excellent concurrence with the Rashba type band superimposed from DFT. Thus, one can conclude that the ARPES data at $hv = 22$ eV are dominated by these bands, while the Dirac cone is barely visible. Beside the specific symmetry of a band, the incident photon energy also plays an important role whether the matrix elements of a band allow a transition from an initial to a final state, which then leads to the detection of the band in the ARPES experiment (cf. eq. 3.21 in section 3.2).

By having a closer look at the Rashba state in the ARPES experiment, it becomes apparent that the energy and momentum resolution is not sufficient to resolve the different spin branches. However the two branches as well as the energy width of the spin splitting between the two bands ΔE_{SO} get visible by mapping spin resolved EDCs at two distinct $\mathbf{k}_{\parallel}$-values (as marked by dashed lines in Fig. 4.15 a)). The corresponding EDCs for the spin direction perpendicular to $\mathbf{k}_{\parallel}$ are displayed in Fig. 4.15 b). The spin-up and spin-down component are again colored in blue and red, respectively, using the same color code as in the DFT calculation. Obviously, there is a spin splitting between the two components which can be quantitatively determined by fitting the peak in each curve by a Lorentzian function as shown by the solid lines. Spin splitting energies ΔE_{SO} which are of the order of 100 meV are found and reasonably agree with the values found in the DFT. The comparison of spin splitting energies between experiment and theory for different $\mathbf{k}_{\parallel}$ is plotted in Fig. 4.15 c). Moreover, as expected for a Rashba-type spin splitting, the spin direction for the upper and lower peak inverts by inverting the $\mathbf{k}_{\parallel}$ direction. The spin direction parallel to $\mathbf{k}_{\parallel}$ was further checked but only negligible spin polarization could be found. This implies that for the in-plane component, the spin of the upper (lower) band rotates clockwise (counter-clockwise) with

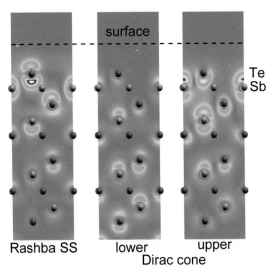

Figure 4.16: Two-dimensional cut through the calculated local density of states for the Rashba state and the lower and upper part of the Dirac cone at $k_{\parallel} = 0.06\,\text{Å}^{-1}$. (Calculation by Gustav Bihlmayer).

respect to $\mathbf{k}_\parallel$, which is in agreement with the spin direction in the DFT calculation. As depicted in Fig. 4.14, DFT further reveals that the Rashba state shows no pronounced out-of-plane polarization within the first two atomic layers whereas in the in-plane direction, the different spin branches are nearly fully polarized. From $k_\parallel = 0.05\,\text{Å}^{-1}$ to $k_\parallel = 0.15\,\text{Å}^{-1}$ the lower spin branch shows a slightly lower in-plane polarization than the upper spin branch. This is probably due to its proximity to the bulk bands. Although there is a small reduction of the polarization, the coupling to the bulk band seems to be rather low. Otherwise, the reduction should be more pronounced. Compared to the DFT, the experimental spin resolved data reveal a net spin polarization of $P_y \simeq 45\,\%$ for the upper band and $P_y \simeq 18\,\%$ for the lower band (Fig. 4.15 b)). Most likely, the overlap of the two bands in the experiment as well as their overlap with the bulk bands caused by the limited energy and momentum resolution of the spin detector is responsible for the small numbers. Moreover, the fact that the peak at higher energy is sharper is probably related to its larger separation from the bulk bands (see DFT in Fig. 4.8) leading to longer lifetime.

The symmetry as well as the origin of the SO generated bands referred to their position in the crystal structure is plotted for $k_\parallel = 0.06\,\text{Å}^{-1}$ in Fig. 4.16. It shows the particular charge density of states for the Rashba state and the lower and upper Dirac cone within the first two QLs. Interestingly, the Rashba SS exhibit predominantly Te p_z character and is localized strongly within the Te top surface layer. In contrast, the states of the Dirac cone are more Sb p_z like and penetrating more strongly into the bulk of Sb$_2$Te$_3$. The different penetration depth also might be a reason why the Dirac cone is more easily observed at higher photon energy, while the Rashba state dominates the spectra at $h\nu = 22$ eV.

From the charge density of states, the electric field between the surface Te-layer and the subsurface Sb-layer can be deduced from the calculated surface core level shift and is found to be about $2 \cdot 10^8$ V/m. This corresponds to a strong dipolar contribution between Te$^{\delta-}$ and Sb$^{\delta+}$, which is probably responsible for the relatively large Rashba coefficient of $\alpha \simeq 1.4\,\text{eVÅ}$, similar to the findings in surface alloys [25] and layered bulk compounds [160].

Rashba spin-split surface state protected by a spin orbit gap

As already briefly mentioned before, the Rashba state present in Sb$_2$Te$_3$ behaves differently from the Rashba bands found so far [22, 23], as the different spin branches disperse into different projected bulk continuum bands (cf. DFT in Fig. 4.8). Thus, each spin branch connects the upper and the lower bulk band surrounding the gap by dispersing from $k_\parallel = -0.28\,\text{Å}^{-1}$ to

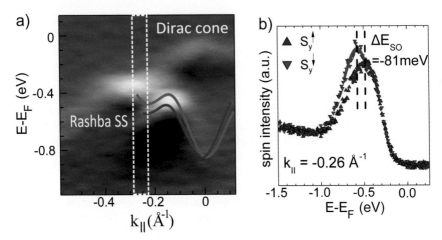

Figure 4.17: a) ARPES data at $hv = 54.5\,\text{eV}$ including the Rashba state at higher $|\mathbf{k}_{\parallel}|$. For better visibility, the derivative with respect to E is shown. The Dirac cone and Rashba SS with superimposed $E(\mathbf{k}_{\parallel})$ curves of the Rashba states (colored lines) from DFT calculation are marked. White dashed box indicate the same area as the dashed box in Fig. 4.18 as well as the $|\mathbf{k}_{\parallel}|$-position of the EDC in (b). The width is given by the angular resolution of the experiment. b) Spin resolved EDC measured at the position of the SO gap ($|\mathbf{k}_{\parallel}|$ as marked); $hv = 54.5\,\text{eV}$; peak positions as determined from Lorentzian fits (solid lines) are indicated by dashed lines; the resulting ΔE_{SO} is marked.

$k_{\parallel} = 0.28\,\text{Å}^{-1}$. At $\overline{\Gamma}$, the surface state is spin degenerate as requested by time reversal symmetry.

Figure 4.17 a) shows the measured band structure close to the point where the Rashba bands merge with the bulk bands according to DFT. Indeed, a band moving upwards and a band moving downwards are discernible up to about $|\mathbf{k}_{\parallel}| = 0.27\,\text{Å}^{-1}$ following the course of the overlaying Rashba bands of the DFT (red and blue lines). The measured spin-resolved EDC, recorded at $|\mathbf{k}_{\parallel}| = 0.26\,\text{Å}^{-1}$ (dashed box in Fig. 4.17 a)), which is close to the point where merging of surface bands and bulk bands is obtained in the calculation, reveals that a spin splitting of about 81 meV is indeed still visible (Fig. 4.17 b)). Thus, this result confirms the DFT calculation which foresees a spin splitting of 80 meV. In order to explain this remarkable behavior, one can have a closer look at the bulk band structure in this particular energy and momentum region. Again, SO interaction is the driving force as becomes obvious from the band structure in Fig. 4.18 which is plotted with (gray background and dashed black lines) and without (yellow background and solid black lines) SO interaction. Obviously the SO interaction opens a gap (SO gap) between

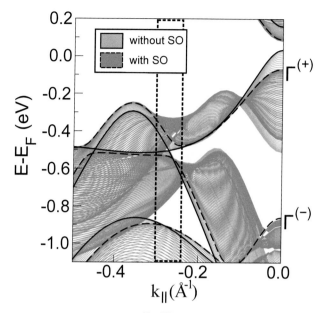

Figure 4.18: Bulk band structure along $\overline{\Gamma} - \overline{K}$ calculated with (gray lines) and without (yellow lines) SO interaction. Black lines mark the band edges with (dashed) and without (solid) SO interaction. In the calculation with SO interaction, a gap opens near the center of the graph around $E = -0.5\,\text{eV}$ and $k_{\parallel} = \pm 0.26\,\text{Å}^{-1}$. $\Gamma^{(-)}$, $\Gamma^{(+)}$ mark the different parity of the two bands at the Γ-point. The black dashed box indicate the same area as the dashed box in Fig. 4.17 as well as the $|k_{\parallel}|$-position of the EDC in Fig. 4.17 b). (Calculation by Gustav Bihlmayer).

the projected bulk states originating from a band $\Gamma^{(+)}$ near the Fermi level and a lower-lying $\Gamma^{(-)}$ band, where $(+)$ and $(-)$ marks the parity of the states at $\overline{\Gamma}$. The gap is found at $k_{\parallel} = 0.26\,\text{Å}^{-1}$ and $E - E_F = -0.5\,\text{eV}$ along the line $\Gamma - \Sigma$ of the bulk band structure ($\overline{\Gamma} - \overline{K}$ in terms of surface Brillouin zone). So apparently, a similar situation as in the bulk band gap at E_F, where the non-trivial nature of the gap is due to the SO interaction inducing an inversion of bands with different parities, is generated here. So the question arises whether this SO generated gap, which is accompanied with a change in parity, also exhibits some non-trivial properties, even if the gap is away from a high symmetry point. In order to analyze this, I go back to a theoretical argument given by Pendry and Gurman [161] in 1975. Independent of topological considerations and only based on numerical calculations of incident and reflected Bloch waves, they provided general criteria for the presence of

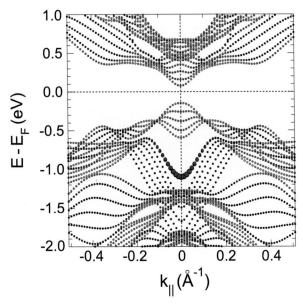

Figure 4.19: DFT band structure calculated for a slab without SO coupling. The surface state (red dots) still exists, but is only connected to the lower bulk band exhibiting $\Gamma^{(-)}$ character. The topological Dirac cone has vanished. (Calculation by Gustav Bihlmayer).

surface states. One major criteria, which is important to our case, is that if there is a SO generated gap present and not located at a high symmetry point of the Brillouin zone, then there must necessarily exist at least one surface state within that gap. In other words, if there is such a gap there must also be a state. Thus, the observed Rashba split surface state, which was experimentally resolved within the SO gap, must necessarily be there, i.e. it is protected by this gap. Hence, even if the argument of the surface state protection by Pendry and Gurman is also only based on bulk considerations, it is different from the topological aspect as the parities of the bands obviously do not play a role. Moreover from this Pendry and Gurman argument, the protection of the Rashba state is only valid in the small area of the SO gap. Nonetheless, a similar behavior to the topological Dirac cone is visible throughout the whole gap-like area (from $k_{\parallel} = -0.27\,\text{Å}^{-1}$ to $k_{\parallel} = 0.27\,\text{Å}^{-1}$), as it connects the lower and the upper bulk bands and is thus energetically present throughout the whole area.

So far, there has been very little experimental proof of such a surface state within a SO gap away from a high symmetry point. Feder and Sturm [162] report of a spin-orbit generated gap along the symmetry line $\overline{\Gamma H}$ in W(001) by means of tight-binding calculation, in which a surface state is found experimentally [163]. Another example is a well defined surface state located in the $\overline{\Gamma} - \overline{T}$ direction in Bi(111) [24, 164]. First principle calculations by Gonze *et al.* [165] resolved a SO gap on this symmetry line which is consistent in energy with the measured surface state. Differently from these states however, our observation reveals a *spin-splitted* surface state with topological character, which, moreover exists in parallel to a surface Dirac cone in the fundamental gap. Thus, it adds a distinct example to Pendry and Gurman's criteria.

The importance of the SO interaction for the unconventional behavior of the Rashba SS is displayed in Fig. 4.19 which shows the band structure of Sb_2Te_3 from DFT without SO interaction. The Rashba state becomes a spin degenerate state and merges for positive and negative momenta with the same bulk band. Hence, it does not connect the upper and the lower bulk band anymore and it loses its protected character, however, it is still present. This proves that the unconventional behavior towards higher wavenumbers is only driven by SO coupling in line with Pendry and Gurman's argument [161]. The Dirac cone at the Fermi level, however, does not exist in this case as the non-trivial nature of the gap is lifted, confirming the topological nature of the surface Dirac cone.

Effective spin polarization of the lower Dirac cone deduced from experiment

Here, I come back to the above described discrepancy between the experimental and theoretical detected spin polarization of the TSS. Whereas calculation predicts an in-plane spin polarization of 90 % for the lower part of the Dirac cone at $k_{\parallel} = 0.06\,\text{Å}^{-1}$ (Fig. 4.14), only a value of $P_y \simeq 20$ % was detected in the experiment. In most experimental works, the reduction of the spin polarization is due to extrinsic factors, like the insufficient instrumental resolution in spin-ARPES measurements especially in the case where different states are relatively close to each other [66, 69]. In our case, the unpolarized background from the bulk valence band (BVB) states together with the low resolution of the spin-ARPES technique considerably reduces the spin polarization. Besides the background from the bulk bands, the spin-resolved spectra feature a finite background from the Rashba-type surface state visible as increasing intensity towards higher binding energies in the spin-resolved spectra in Fig. 4.12 a) and b). After subtraction of this background by using a Gaussian function, the spin polarization already slightly increases to a value of $P_y \simeq 27$ % for $k_{\parallel} = -0.06\,\text{Å}^{-1}$ (Fig. 4.20). However, the major reduction must thus originate from the unpolarized bulk bands.

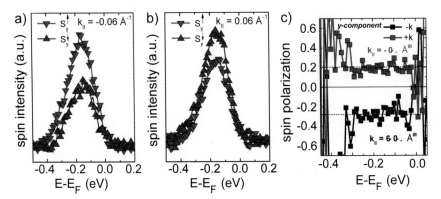

Figure 4.20: a), b) Spin-resolved EDCs for the spin component perpendicular to $k_\parallel$ after subtraction of the Rashba state background (compare with Fig. 4.12 a) and b)). c) Corresponding spin polarization as a function of energy deduced according eq. 4.2. The subtraction of the background increases the averaged spin polarization from 20 % to 27 %.

In this way, it is essential to evaluate the finite contribution from the spin-degenerate BVB close to the lower Dirac cone to the ARPES intensity [166] (Fig. 4.8 and Fig. 4.11 b)) in order to discuss the absolute spin polarization more quantitatively. This is easiest achievable in the high-resolution ARPES data (Fig. 4.11 b)), as there, one is able to clearly distinguish between the Dirac cone peak and the BVB background. However, in order to do so, one must firstly show that the measurements taken with the high-resolution apparatus[3] are comparable with the data achieved by the spin-resolving detector[4].

In Fig. 4.21 a) the close-up ARPES measurement of the Dirac cone recorded with high-resolution is shown. The Dirac cone and the adjacent BVB are visible. If this data is convoluted with a two-dimensional Gaussian function having the energy and momentum resolution of the spin-resolving detector as FWHM in the two directions, the Dirac cone part strongly overlaps with the BVB (cf. Fig. 4.21 b)). Indeed, a similar broadening of the structure is found in the ARPES data recorded with the spin detector (Fig. 4.21 c)), albeit with much weaker intensity due to the lower efficiency of the spin resolving apparatus. Moreover, the corresponding constant-energy cuts from both the convoluted ARPES data (Fig. 4.21 e)) and the spin-ARPES data (Fig. 4.21 f)) exhibit a similar distribution. Thus, both the data from the spin-ARPES (Fig. 4.21 c)) and the high-resolution ARPES (Fig. 4.21 a)) only differ by their differ-

---

[3]The energy and angular resolution are 20 meV and 0.2 °, respectively.
[4]The energy and angular resolution are 100 meV and 1.4 °, respectively.

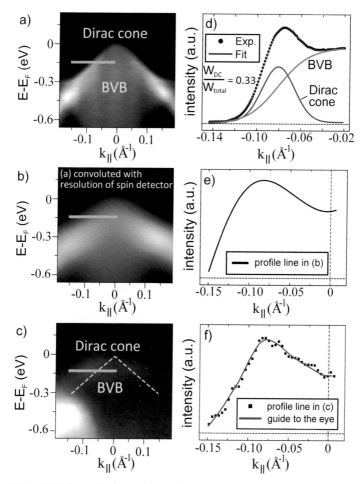

Figure 4.21: a) High resolution ARPES data of the Dirac cone (close-up view from Fig. 4.11 b)) at an incident photon energy $h\nu = 55$ eV. b) data from (a) convoluted with a Gaussian curve taking into account the energy resolution (FWHM: 100 meV) and momentum resolution (FWHM: 0.09 Å^{-1}) of the spin detector. c) ARPES data of the Dirac cone measured with the spin detector, $h\nu = 54.5$ eV. d)−f) Constant-energy cuts through the Dirac cone along the green line in the ARPES data aside. Constant energy cuts in (e) and (f) show a similar distribution. In (d), the result of the two component fitting to the spectrum is shown as marked Dirac cone and BVB. Ratio of the spectral weight of the Dirac cone W_{DC} to the total spectral weight W_{total} is indicated.

ent resolution, such that one can consider the data from Fig. 4.21 a) in order to deduce the contribution from the BVB in the spin polarization of the TSS.

The quantitative analysis is shown in Fig. 4.21 d) highlighting the contributions of the Dirac cone and the BVB as fitting curves. I employed a Lorentzian function for the Dirac cone and a tanh for the BVB. In the case of the Dirac cone, the resulting peak is wider than the momentum resolution of the high resolution detector. Thus, assuming a Lorentzian function, which takes into account the lifetime broadening of the surface state, is justified. The BVB contribution is approximated by a tanh function implying a BVB of constant density within the Dirac cone as found in the DFT calculations and with no bulk bands beyond the Dirac cone. The latter implies the additional restriction within the fit that the reversal point of the tanh function matches the peak position of the Lorentzian. Width and height of the two functions are used as independent fitting parameters.

From the two fit curves, the spectral weight of the Dirac cone states with respect to the total spectral weight within the spin-ARPES measurement is estimated. Therefore, one determines the area of each fit curve in Fig. 4.21 d) within the width of the angular resolution of the spin-ARPES experiment of $0.09\,\text{Å}^{-1}$ around the probed $k_{\parallel}$-value of $0.06\,\text{Å}^{-1}$. The weights are

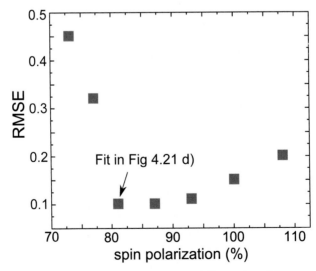

Figure 4.22: Root-mean-square error (RMSE) for different fits of the spectrum of Fig. 4.21 d) leading to different W_{DC}/W_{total} and, thus, to different spin polarization values of the Dirac cone states P_{DC}. The RMSE increases by a factor of three for a spin polarization below 80 %.

called W_{DC} and W_{BVB}, respectively. I checked that the energy resolution of the experiment is not relevant for the angular broadening. The result is $W_{DC}/(W_{DC} + W_{BVB}) = 0.33$ as indicated in the plot.

The resulting intrinsic spin polarization of the Dirac cone in-plane component is then given by the experimentally measured spin polarization P_y divided by W_{DC}/W_{total} leading to a value of $P_{DC} \simeq 82\,\%$. To check the accuracy of the employed fitting procedure, the development of the root-mean square error (RMSE) of the fitting was analyzed for fixed different relative heights of the Lorentzian and the tanh function, which is equivalent to the outcome of different P_{DC}. The RMSE as a function of P_{DC} is plotted in Fig. 4.22. The data reveals that a fitting procedure with a reasonable RMSE ends with a spin polarization of 80-95 % for the Dirac cone states, so that one can conclude that the real spin polarization is likely to be within that range. Notice that this result also nicely agrees with the DFT result of $P_{DC} = 90\,\%$. Moreover, the energetic width of the Dirac cone peak in spin-ARPES (Fig. 4.20) is similar to the total width of the BVB in the DFT calculation (Fig. 4.8), which supports that our estimate of P_{DC} is reasonable, i.e. that the BVB contribution leads to a peak of similar width as the Dirac cone peak. In turn, high-resolution spin-ARPES experiments are required to measure the intrinsic spin polarization of the Dirac cone directly, i.e., without relying on any assumption.

Finally, in line with the STS data, spin-ARPES revealed the topological nature of the Dirac cone within the fundamental gap of Sb$_2$Te$_3$, and characterized the spin texture of the surface state, which rotates counter-clockwise for the lower part of the Dirac cone. A spin polarization of up to 90 % could be detected. In addition, in accordance with DFT calculations, a novel, strongly spin-split Rashba-type surface state was identified which is protected by a SO gap away from $\overline{\Gamma}$ and connects an upper and a lower bulk valence band. This state is similarly to the TI state protected by symmetry according to a fundamental criterion given by Pendry and Gurman in 1975.

4.3 Characterization of crystalline Ge$_2$Sb$_2$Te$_5$ in terms of topological insulator

The last section dealt with the TI properties of crystalline Sb$_2$Te$_3$ which is at the border of the so-called pseudobinary line (cf. Fig. 4.1). In the present section, another PCM of the pseudobinary line will be analyzed, namely the ternary compound GST-225, a prototype PCM, which is also predicted to be a promising candidate for TI behavior [120, 123]. However, the question if GST-225 is a TI is closely related to its stacking sequence [123]. Most of the results of this section are published in ref. [30].

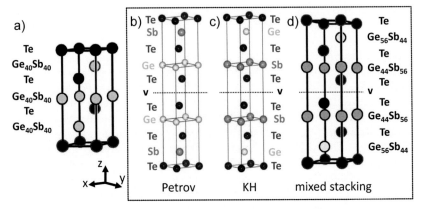

Figure 4.23: a) Cubic rocksalt structure of the metastable GST-225 phase along the [111] direction. b)-d) Three possible stacking sequences for the crystalline hexagonal stable phase along [0001]. Ge, Sb, Te and vacancies (v) as marked. ((a) and (d) from ref. [138]; (b) and (c) from ref. [120]).

4.3.1 Stacking order of $Ge_2Sb_2Te_5$ in the crystalline phase

GST-225 emerges in two slightly different crystalline phases, i.e. a metastable cubic one used for applications [131] and a stable hexagonal one, however the stacking sequence of both phases is yet not fully determined. Figure 4.23 shows the cubic rocksalt structure for the metastable GST-225 phase, along with three possible stacking sequences for the hexagonal phase. For means of comparison with the stable hexagonal phase, the rock salt structure is displayed along [111] exhibiting hexagonal layers with $(Te-Ge/Sb/v)_3$ sequence, where $Ge/Sb/v$ is a mixed layer of Ge, Sb and vacancies [138, 167] (Fig. 4.23 a)). Figure 4.23 b)-d) shows three possible stacking orders for the hexagonal phase which are built along [0001] and mainly differ on their respective Ge and Sb layers. The sequences in b) and c) are deduced from transmission electron microscopy (TEM) and are either Te-Sb-Te-Ge-Te-v-Te-Ge-Te-Sb- (Petrov phase) [139] or Te-Ge-Te-Sb-Te-v-Te-Sb-Te-Ge- (Kooi-De Hosson or KH phase) [140]. The v denotes a vacancy layer where adjacent Te layers are van-der-Waals bonded. In both stackings, each layer is a pure layer of only one particular element and no mixture of layers as in the metastable rocksalt structure takes place. DFT calculations imply that the KH phase is slightly energetically favorable with respect to the Petrov phase [168]. However, more recent X-ray diffraction data suggest that some mixture of Ge and Sb in the respective layers is present (Fig. 4.23 d)) [138]. Thus, the distribution of the Ge and Sb within the layers is still under discussion and it turns out that the exact occupation is a critical point for a possible TI nature of the GST-225 compound. For the metastable rocksalt structure, TEM studies suggest that the Ge/Sb/v layers exhibit some internal order [169] and DFT even implies that Ge, Sb and vacancies might accumulate in separate layers [168]. From this perspective, the stable and the metastable phase could be much closer than originally anticipated and the transition between them would be a mere shift of blocks of (111) layers without atomic rearrangements within the layers [168].

So far, there have been no calculations including SO interaction for the metastable rocksalt phase, in contrast to the stable hexagonal phase which is easier to simulate as the structure is much more regular. In the metastable phase, the vacancies are randomly distributed whereas in the hexagonal phase they occupy pure single layers, which is preferred for DFT. The first prediction of topologically insulating GST-225 was made by Kim *et al.* for the Petrov phase while the energetically favorable KH phase was shown to be topologically trivial [120]. The KH phase can be considered as a short-period superlattice consisting of the TI Sb_2Te_3 and the trivial band insulator GeTe [121]. In this aproach, Kim *et al.* [121] revealed that a transition from a trivial to a non-trivial phase occurs when the ratio of Sb_2Te_3 relative to that of the insulating GeTe increases in the GST compound. Nonetheless, DFT showed that even a GST composition which is in a trivial phase, e.g. the KH phase of GST-225, can be transformed into a TI if the material is set under isotropic pressure [170] or constant strain [171]. A transition from the Petrov to the KH phase with a more disordered mixed-layer phase (structure as in Fig. 4.23 d)) in between has been investigated by Silkin *et al.* by DFT [123]. They considered a stacking sequence of Te-M1-Te-M2-Te-v-Te-M2-Te-M1- with a tunable Ge/Sb ratio for the layers M1 and M2 ($Ge_{2x}Sb_{2(1-x)}$ in M1 and $Ge_{2(1-x)}Sb_{2x}$ in M2) and determined the $\mathbb{Z}_2$ topological invariant ν_0 for well defined x. The Petrov phase ($x = 0$) was found to be a semimetal and the KH phase ($x = 1$) a trivial insulator, however in the mixed phase, a TI nature was observed for $x = 0.25$ and $x = 0.5$. Thus, the value of ν_0 crucially depends on the respective Ge concentration within the mixed layer. Further, this result suggest that in a real material, disorder within the layers will not prevent the existence of a topological surface state [123]. Note, that the ν_0 invariant could not be calculated for the Petrov phase ($x = 0$) due to the lack of a real band gap in the band structure.

As already mentioned above, there are no calculations of topological properties for the more disordered metastable rocksalt phase in the literature so far. Thus, the results of the more disordered hexagonal phase by Silkin *et al.* are closest to the experimentally studied metastable phase. Importantly, all DFT calculations of GST-225 in the literature so far exhibit the valence band maximum (VBM) away from Γ when revealed in the topologically non-trivial

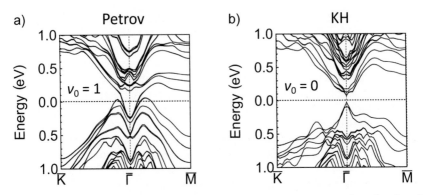

Figure 4.24: Calculated surface band structure including SO coupling for the a) Petrov sequence and b) KH sequence with the $\mathbb{Z}_2$ topological invariant v_0 as marked. In the non-trivial case (a), the valence band maximum (VBM) is away from the Γ point, whereas in the trivial case (b) VBM is located at Γ. (Adopted from [120]).

phase [120, 121, 170, 171, 123]. However if the GST-225 compound is found to be in the trivial phase (e.g. KH phase or $x = 0.75$ and $x = 1$ in ref. [123]) the corresponding band structure always exhibit a VBM exactly at Γ. Both, the surface band structure including SO for the non-trivial Petrov and the trivial KH phase of GST-225 from ref. [120] are displayed in Fig. 4.24 demonstrating the described VBM character at Γ for the respective v_0 number. Thus, the noted calculation all show the same clear tendency that for GST-225 the topmost valence band course is directly connected to the topological invariant. In the following, the differentiation of the topmost valence band course will be taken as the central argument for the characterization of the experimental data.

4.3.2 Preparation and characterization of metastable crystalline Ge$_2$Sb$_2$Te$_5$ samples

In order to study TI properties by ARPES, ideally single crystalline GST samples are desired. Typically, however, GST is sputter-deposited resulting in polycrystalline films. Only recently, epitaxial films of superior crystalline quality have been grown by molecular beam epitaxy (MBE) on GaSb, InAs, and Si [172, 173, 174, 175, 176]. The metastable cubic, rhombohedrally distorted GST-225 grows with a single vertical epitaxial orientation, well-defined interfaces, and atomically flat terraces only on (111)-oriented substrates [172, 173].

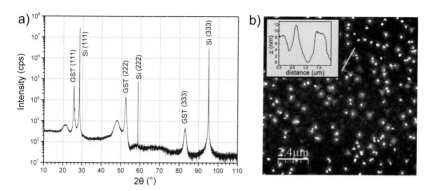

Figure 4.25: a) XRD measurement of the epitaxially grown GST-225 sample of 20 nm thickness on a Si(111) substrate. The data reveals the metastable cubic phase along the [111] direction. The respective peaks of GST-225 and the underlying Si(111) substrate are labeled by their Miller indices, respectively. The additional peaks (not labeled) point to the presence of a superstructure. (XRD data recorded by Alessandro Giussani). b) Tapping mode AFM under ambient conditions revealing the typical surface appearance of a MBE grown GST-225 sample with a roughness ≈ 5 nm (RMS). Inset shows the height profile of the area marked by the straight line.

The GST-225 thin films (thickness 20-30 nm) measured in this work have been grown by MBE on a Si(111) substrate by the group of Dr. Raffaella Calarco at the Paul-Drude Institut in Berlin. The temperature of the effusion cells in the MBE chamber was set to $T = 250\,°C$ so that the flux ratio of Ge:Sb:Te is close to 2:2:5, as has been confirmed by X-ray fluorescence [172, 173]. After preparation, XRD measurements have been used to confirm the metastable cubic phase along the [111] direction of the GST films (Fig. 4.25 a)). The Bragg peaks of cubic GST-225 and of the underlying Si(111) substrate have been detected. The presence of superstructure peaks in addition to the Bragg reflections indicates a vacancy ordering in the Ge/Sb/v sublattice along the growth direction [177]. Additional atomic force microscopy (AFM) measurements show the large-scale topography of the GST-225 surface with flat terraces of widths of several 100 nm (Fig. 4.25 b)). A roughness of ≈ 5 nm (RMS) was found.

After growth, the samples have been transferred under ambient conditions. Differently to the Sb$_2$Te$_3$ single crystal samples, a simple cleavage of the GST-225 films prior to the ARPES measurements was not possible due to the small thickness of the films and the stronger bonding between adjacent layers in the [111] direction perpendicular to the surface. Hence, a different method which effectively removes any kind of native oxides and adsorbants from the surface has been applied. Zhang *et al.* [178] proposed a technique which implies the

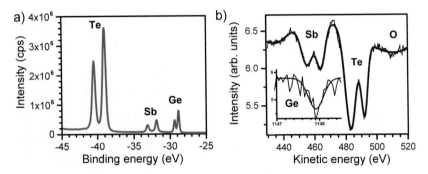

Figure 4.26: GST-225 sample after DI water dip and UHV annealing. a) XPS at $h\nu =$ 110 eV revealing the Ge 3d, Sb 4d and Te 4d levels. b) Auger electron spectroscopy (AES) revealing the peak positions of Ge, Sb, Te and O as marked. The straight lines are Gaussian fits used to determine the stoichiometry to be: 18 % Ge, 21 % Sb, 57 % Te and 4 % O. (Data by Jens Kellner [179]).

dipping of the GST-225 film into de-ionized (DI) water for 1 min, followed by an immediate transfer of the de-oxidized sample into the UHV chamber. The DI-water tackles above all the Ge- and Sb-oxide bondings so that a deficiency of Ge and Sb is left after the cleaning. Moreover, after the dipping, the DI-water remains on the surface and acts as a protecting film for several minutes so that no new contamination of the surface takes place within this time. In the UHV chamber, the de-oxidized sample is annealed to 250 °C so that Ge and Sb can diffuse from the bulk to the surface and restore the stoichiometry of 2:2:5. Zhang *et al.* confirmed the accuracy of the dipping method, namely the cleanliness of the sample as well as the recovery of the stoichiometry, by means of XPS, AFM and XRD [178].

Here in this work, XPS is used prior to the ARPES measurements in order to verify the cleanliness of the sample surface after the DI water dip process. And indeed, clear peaks were observed in the XPS spectrum which could be assigned to the Ge 3d, Sb 4d and Te 4d levels [180, 181] as marked in Fig. 4.26 a). The peaks further showed no sign of distortion, implying a clean and oxygen-free surface. The stoichiometry of the sample after the cleaning process has been checked by means of Auger electron spectroscopy (AES)[5] (Fig. 4.26 b)). The AES spectrum shows the peak position of the GST elements as marked, as well as a small peak arising from remaining oxygen left on

---

[5]AES is based on the Auger effect which includes the emission of electrons from an excited atom after a series of internal collision events. The energy of the emitted electron is element-specific and provides information about the composition of the sample. Typical excitation energies are in the order of 1-3 keV.

the surface. The stoichiometry has been calculated from the respective peak intensities for each element [182] using the tabulated sensitivities for Ge, Sb, Te and O. An amount of 57 % Te, larger than its initial stoichiometric part, has been detected and is partly attributed to the Te termination of the surface leading to larger AES intensities. Ge (18 %) and Sb (21 %) further confirm the recovery of the stoichiometry after cleaning. Most notably, the oxygen content of the surface is only 4 % [179].

We further checked that neither the measurement nor the preparation process lead to an unintended phase transition from the metastable cubic phase into the stable hexagonal phase. This transition is expected to take place at $\approx 340\,°C$ [183]. For that purpose, XRD has been performed after the ARPES measurements, revealing the same cubic structure with vacancy ordering as was observed directly after the MBE growth. Note that a further heating of the sample in UHV at $\approx 300\,°C$ in order to create a phase transition was found to cause considerable change in stoichiometry due to the different desorption temperatures of atomic species. At the same time, a change in the peak structures of the Ge 3d and Sb and Te 4d levels in XPS was observed.

For a characterization of the epitaxially grown GST-225 samples, the topography of the cleaned surface has been investigated at room temperature by STM on the nm-scale [179]. Atomically flat terraces of up to 100 nm in width are found (Fig. 4.27 a)). These terraces are separated by steps of $\approx 0.34\,nm$ in height, which corresponds to the expected Te-Te layer distance of 0.347 nm in the [111] direction of cubic GST-225 [184, 185]. On the terraces, atomic resolution is achieved showing a hexagonal appearance (Fig. 4.27 c) and d)), most likely originating from the Te layer [185]. The atomic distance is found to be 0.43 nm which nicely agrees with the expected atomic distance for the (111) surface of the cubic phase ($a = 0.42\,nm$) [184, 185]. Additionally, $dI/dV(V)$ spectra have been recorded showing the electronic structure of GST-225. A band gap of $\approx 0.4\,eV$ with E_F situated at the top of the valence band is resolved (Fig. 4.27 b)), also indicating a strongly p-type nature of the material, probably due to the large amount of vacancies.

4.3.3 Evidence for topological band inversion in metastable crystalline Ge₂Sb₂Te₅ measured by ARPES

Similar to the Sb₂Te₃ data, the ARPES measurements on the epitaxially grown GST-225 has been preformed at the beamline of the synchrotron BESSY in Berlin. Figure 4.28 a) displays a spectrum recorded with linearly polarized light at $h\nu = 22\,eV$ in a direction determined to be $\bar{\Gamma} - \bar{K}$ by comparison with DFT calculations. Just below E_F, the upper valence band shows maxima at $k_\parallel = \pm 0.14 \pm 0.02\,\text{Å}^{-1}$ and drops to $E - E_F = -0.3\,eV$ at $\bar{\Gamma}$. Another band re-

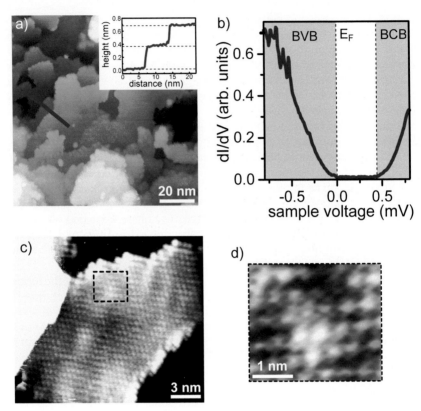

Figure 4.27: a) STM image of the cleaned GST-225 (111) (V = -0.3 V, I = 100 pA) at room temperature reveals atomically flat terraces. Blue line marks the position of the height profile in the inset. Average step height of ≈ 0.34 nm is observed. b) $dI/dV(V)$ curve ($V_{stab} = -0.8$ V, $I_{stab} = 100$ pA, $V_{mod} = 8$ mV) of GST-225 (average of 10 spectra) shows a band gap of ≈ 0.4 eV. Gray shaded areas mark the bulk valence (BVB) and bulk conduction band (BCB), respectively. c) STM image with atomic resolution (V = -0.5 V, I = 100 pA). d) Zoom into the area marked by a dashed box in in (c). A hexagonally arranged pattern, most likely of the Te atoms is observed with an average atomic distance of 0.43 nm. A triangular defect similar to the Sb_{Te} antisite defect in Fig. 4.7 b) is visible. (Data by Jens Kellner [179]).

sides between -0.7 eV at $k_{||} = \pm 0.23$ Å^{-1} and -0.35 eV at $k_{||} = \pm 0.1$ Å^{-1}. Closer to $\bar{\Gamma}$, these two bands lead to a broad peak in energy distribution curves (EDCs) around -0.4 eV with a FWHM of 0.5 eV (Fig. 4.30 b)). Below -1 eV, there are two more hole-like bands. Figure 4.28 b) shows a constant energy cut of the ARPES data at E_F displaying a nearly isotropic behavior of

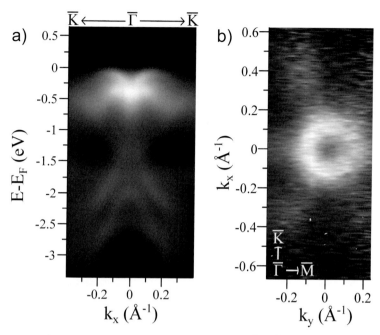

Figure 4.28: a) ARPES spectra at room temperature in $\bar{\Gamma} - \bar{K}$ direction at $h\nu = 22\,\text{eV}$ of metastable cubic GST-225 (111) after DI-water dip (Shirley-type background subtracted). b) Constant energy cuts at E_F with $\bar{\Gamma} - \bar{M}$ being the horizontal direction as marked.

the upper valence band with a clear intensity minimum at the $\bar{\Gamma}$ point. From this plot, the band structure in $\bar{\Gamma} - \bar{M}$ looks essentially the same, however with slightly more intensity at higher $|k|$ values in the six $\bar{\Gamma} - \bar{M}$ directions. Fig. 4.29 a) shows a close-up view of the upper band around the $\bar{\Gamma}$-point and corresponding constant energy cuts (Fig. 4.29 b)-d)) at energies marked by the dotted lines. A sixfold symmetry of the valence band at larger $k_{||}$ values is visible with a clear distinction between the $\bar{\Gamma} - \bar{K}$ and $\bar{\Gamma} - \bar{M}$ direction. Since DFT calculations of cubic metastable GST-225 do not show any bands at higher k than $0.3\,\text{Å}^{-1}$ down to $-0.2\,\text{eV}$ in $\bar{\Gamma} - \bar{K}$ direction, but bands at such high k values in $\bar{\Gamma} - \bar{M}$ (see Fig. 4.31), the direction with intensity at high k values in Fig. 4.28 b) has been attributed to the $\bar{\Gamma} - \bar{M}$ direction.

In order to distinguish between surface bands and bulk bands, energy dispersions at different photon energies (Fig. 4.30 a)), which is equivalent to different k_z-values ($h\nu = 17 - 26\,\text{eV}$) in the Brillouin zone, has been probed. The upper valence band changes with photon energy revealing itself as a bulk

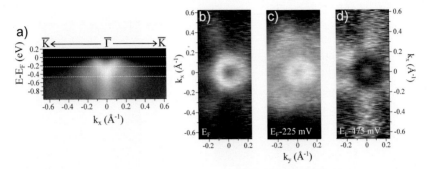

Figure 4.29: a) Close-up view of the band structure at E_F in $\bar{\Gamma} - \bar{K}$ direction at $h\nu =$ 22 eV. Dotted lines mark the energies of the constant energy cuts in (b)-(d). b)-d) Constant energy cuts in $k_{||}$-directions at energies as indicated. A sixfold symmetry is visible at all energies displayed. Note the rotation of the star-like structure, i.e. the maximum intensity, between (b) and (d) by 30°.

band, while the two bands below -1 eV do not. The corresponding EDCs at the $\bar{\Gamma}$-point and near the maximum of the highest band ($k_{||} = 0.12\,\text{Å}^{-1}$) which are shown in Fig. 4.30 b) and c), respectively, confirm this tendency. Both plots show a dispersive behavior of the upper valence band whereas the maxima below -1 eV show no dispersion in the surface normal direction. The topmost maximum at $\bar{\Gamma}$ shifts down by about 0.2 eV between $h\nu = 22$ eV and 26 eV, indicating a k_z dispersion. One can conclude that the topmost valence band possesses a bulk-like character. Furthermore, the $k_{||}$ position of the valence band maximum (VBM) with respect to photon energy has been analyzed. For determination of the VBM, EDCs for different $k_{||}$ are evaluated. The $k_{||}$ values, for which the valence band peak is highest in energy, is taken as the position of the VBM and defined as $k_{||,max}$. This procedure is applied for the ARPES spectra of 6 different photon energies (Fig. 4.30 a)) and entered into Fig. 4.30 d). Since only EDCs at constant $k_{||}$ are used, variations in ARPES intensity with detection angle, or $k_{||}$, do not influence the outcome. One observes that the VBM also shifts with photon energy revealing a small dependence on k_z as well. Thus, the ARPES peak at the VBM is, at least partially, a bulk band with dispersion in k_z-direction.

In order to analyze the measured ARPES band structure in terms of topological properties, it is useful to compare the ARPES with the STS data, as the $dI/dV(V)$ spectrum also provides information about the unoccupied states. From the STS, it was shown that the Fermi level in the cubic GST-225 films lies just at the edge of the valence band (Fig. 4.27 b)). This implies that the topmost band in the ARPES measurement correspond to the valence band edge. From this knowledge, one can now compare the course of the upper valence band

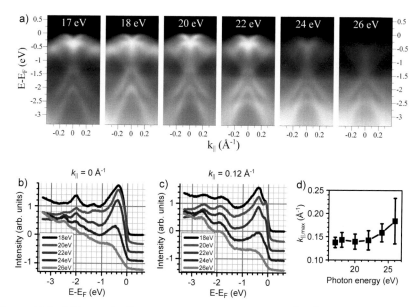

Figure 4.30: a) Electronic dispersion mapped by ARPES in $\bar{\Gamma} - \bar{K}$ direction at different photon energies as marked. b) and c) EDCs for different photon energies (cuts from (a)) at the $\bar{\Gamma}$-point (b) and at $k_{||} = 0.12\,\text{Å}^{-1}$ (c), i.e. near the position of the valence band maximum. Graphs are offset for clarity. d) $k_{||}$-value of the valence band maximum ($\bar{\Gamma} - \bar{K}$ direction) for the different photon energies extracted from (a).

with the corresponding band in the DFT in order to find hints for topological properties. As there have not been any DFT calculations of the metastable cubic phase including SO interaction, a combined bulk and surface calculation for the Petrov and KH stacking is provided here. Similar to the DFT calculations of the phase change compound Sb₂Te₃, the calculations have again been performed by Gustav Bihlmayer within the generalized gradient approximation [151]. The full-potential linearized augmented plane-wave method in bulk and thin-film geometry [186] as implemented in the Fleur code⁶ has been employed. SO coupling was included self-consistently and a basis set cutoff of $R_{\text{MT}}k_{\text{max}} = 9$ was used. As structural model for the cubic phases, the atomic positions given by Sun *et al.* [168] have been adopted both for the bulk and film structures. For the latter, films consisting of 27 atomic layers terminating by a vacancy layer were used. Two different stacking sequences were assumed for the cubic phase: a Petrov- and a KH-like sequence which

---

⁶for a program description, see http://www.flapw.de.

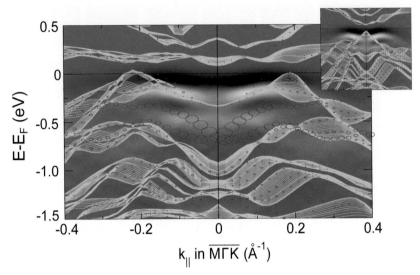

Figure 4.31: DFT calculations of the band structures for cubic GST-225 with Petrov- and KH-type (inset, same scale as main image) stacking sequence, as proposed in ref. [168]. Bulk bands are given as white lines, states of the surface film calculations with circles. The extension of the states into the vacuum (region above the topmost Te layer) is indicated by the size of the circles. The calculations are superimposed with the ARPES spectra (2nd derivative of intensity with respect to electron energy) at 22 eV photon energy. Calculations are shifted upwards by 100 meV. (Calculation by Gustav Bihlmayer).

are derived from the respective hexagonal phases by introducing a shift of one part of the unit cell within the [0001] plane.

The resulting DFT band structure for the Petrov stacking of the cubic metastable GST-225 along with the 2nd derivative of the measured band structure is shown in Fig. 4.31. The inset reveals the same calculation for the KH stacking superimposed on the measured ARPES data. The gray shaded lines in the DFT result from the bulk calculation whereas surface states are marked by red circles. Again, as in the case of the hexagonal phase, the Petrov and the KH sequence predominately differ by the course of the upper valence band. Here, a qualitative agreement with the calculations of the hexagonal phase [74, 123] is found with the KH sequence exhibiting a VBM at Γ for all k_z while the topological Petrov sequence shows the VBM away from Γ for all k_z. A reasonable agreement is obviously only achieved with the Petrov-like stacking, including the minimum at $\bar{\Gamma}$ of the upper valence band. Moreover, in the Petrov stacking, the calculation shows the topological surface state (small red

Table 4.1: $k_{\parallel,max}$ positions of experimental and theoretical valence band maxima given in Å^{-1}, theoretical values from the literature are extracted from graphs in the cited publications using only the topologically non-trivial phases. The percentages (25 %, 50 %) denote the fraction of Ge in the M1 layer.

this work		Kim *et al.*[74]		Silkin *et al.*[123]	
	cubic		Petrov	25%	50%
$\Gamma - K$ (DFT)	0.19	$\Gamma - K$ 0.18		$\Gamma - K$ 0.29	0.30
$\Gamma - M$ (DFT)	0.22	$\Gamma - M$ 0.26		$\Gamma - M$ 0.51	0.52
$h\nu = 20\,\mathrm{eV}$ (exp., $\bar{\Gamma} - \bar{K}$)	0.14	$\bar{\Gamma} - \bar{K}$ 0.18		$A - H$ 0.20	0.16
$h\nu = 26\,\mathrm{eV}$ (exp., $\bar{\Gamma} - \bar{K}$)	0.18	$\bar{\Gamma} - \bar{M}$ 0.21		$A - L$ 0.25	0.21

circles) crossing the Fermi energy in close vicinity of the upper valence band at $k_{\parallel} \approx 0.12\,$Å^{-1}. This state obviously overlaps with the upper bulk valence band within the ARPES data. The topological surface state is necessarily absent for the KH sequence (inset of Fig 4.31). The bands further down in energy (around $-0.6\,$eV at $\bar{\Gamma}$) can be associated with a Rashba-type surface state, similar to the one observed in Sb$_2$Te$_3$ (Fig. 4.8). In comparison to the topological surface state, the Rashba state shows a stronger surface character (cf. size of the red circles in the DFT). In the ARPES data, however, the Rashba-type surface state is not visible which might be due to prohibited transitions for this particular band in the ARPES experiment at the probed photon energy. Similar findings have already been discussed for Sb$_2$Te$_3$ in the previous section.

I finally compare the metastable cubic phase with previous DFT calculations of the very similar hexagonal phase. Most notably, a VBM away from $\bar{\Gamma}$ consistently indicates topologically non-trivial properties for GST-225 [120, 121, 170, 171, 123]. The experimental values of the VBM for the $\bar{\Gamma} - \bar{K}$ direction with respect to calculations of the hexagonal stable phase and the cubic metastable phase are displayed in Table 4.1. The calculated $k_{\parallel,max}$ of the bulk valence band in $\bar{\Gamma} - \bar{K}$ direction ($0.19\,$Å^{-1}) for the metastable cubic phase is slightly larger than the experimental one ($0.14 - 0.18\,$Å^{-1}). This can be explained by the overlap of the bulk valence band with the surface state which crosses E_F at $k_{\parallel} \approx 0.12\,$Å^{-1} and thus shifts the averaged band maximum detected in the experiment towards smaller $k_{\parallel}$-values (Fig. 4.31). Compared to the calculation of the hexagonal stable phase, best agreement with the experiment is found with the slab calculation of the Petrov phase [74] and with the mixed phase with equal distribution of Ge and Sb ($x = 0.5$) [123]. Within the Brillouin zone of this phase, $k_{\parallel,max}$, the valence band maximum projected onto the (0001) plane, is closest to $\bar{\Gamma}$ at the edge of the Brillouin zone in z-direction (see cuts connecting the $H - A - L$ points in ref. [123]). For other phases or mixed distribution ratios, the measured $k_{\parallel,max}$ is always smaller than the calculated $k_{\parallel,max}$ of the bulk VBM of topologically non-trivial hexagonal stable phases of GST-225 ($0.16 - 0.52\,\text{Å}^{-1}$) (cf. Table 4.1).

In-situ transfer of MBE grown $Ge_2Sb_2Te_5$ samples to the ARPES chamber

In order to avoid the DI-water dipping method with a subsequent annealing to 250 °C, a UHV transfer of the epitaxially grown samples to the ARPES chamber has been implemented. A special UHV suitcase has been constructed within our institute by Sven Just, ensuring a base pressure of 10^{-10} mbar during the transfer. The advantage is that no exposition of the freshly prepared sample to ambient conditions takes place. As the previous cleaning method still left some amount of oxygen on the surface sample (cf. Fig. 4.26 b)), the UHV transfer is expected to avoid major contamination, resulting in a better resolution of the ARPES spectrum.

Figure 4.32 shows the corresponding band structure measurements on the in-situ transferred metastable GST-225 thin film. And indeed, a much sharper ARPES dispersion is visible compared to the data in Fig. 4.28. Again, the isotropic behavior of the upper valence band is revealed together with the already observed sixfold symmetry in the constant energy cut. However, no further features are resolved and the general trends of the band structure, as e.g. the position of $k_{\parallel,max}$, are confirmed. For future measurements of samples which can not be cleaved in UHV, this particular UHV transfer will be the method of choice as the quality of the data strongly increases. The sample shown in Fig. 4.32 has been used by J. Kellner *et al.* (to be published) to map

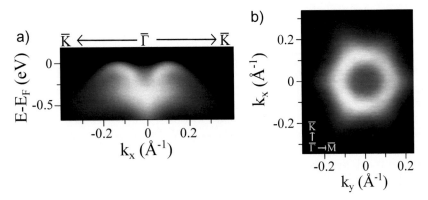

Figure 4.32: a) ARPES spectra at room temperature in $\bar{\Gamma} - \bar{K}$ direction at $h\nu = 22\,\text{eV}$ of metastable cubic GST-225 (111) after in-situ transfer. b) Corresponding constant energy cut at E_F.

the photon energy dependence in more detail, which, e.g. revealed a spin-polarized surface state visible at $h\nu = 50\,\mathrm{eV}$ with helical spin-polarization close to 100%.

In summary, it was shown by ARPES and STS that metastable cubic GST-225 epitaxially grown on Si(111) exhibits valence band maxima $0.14 - 0.18\,\text{Å}^{-1}$ away from $\bar{\Gamma}$ and a band gap of 0.4 eV. All DFT calculations of GST-225 find a VBM away from Γ only for a $\mathbb{Z}_2$ topological invariant $\nu_0 = 1$. This implies topological properties of GST-225, indicating that all phase change materials on the pseudobinary line between Sb$_2$Te$_3$ and GST-225 are topologically non-trivial. Thus, this opens up the possibility of switching between an insulating amorphous phase and a topological phase on ns-time scales. In order to give a direct proof of the topological properties of GST-225, spin-resolved ARPES as in the case of Sb$_2$Te$_3$ is required as has been done recently. Moreover, a better resolution of the ARPES experiment would be helpful in order to distinguish between the different surface states and the bulk valence band. In the $k_{||}$-range of 0 to $0.12\,\text{Å}^{-1}$, the topological surface state lies below the Fermi level and should be detectable in the ARPES experiment. Further, STS at low temperature and in a magnetic field could be an appropriate technique in order to resolve the TI nature of the surface states by Landau level spectroscopy. Then, it has to be proven that the topological properties of the metastable phase survive several switching events, which is not clear a priori, given the fact that the topology seems to depend in detail on the order within the Ge/Sb/v layer.

5 Weak Topological Insulator

Within this chapter, I will switch from the class of strong topological insulators to its counterpart, the class of weak topological insulators (WTIs). As already discussed in the theoretical part of this work (cf. section 2.2.6), the term *weak* is rather misleading as it is referring to the wrong, initial believe that WTIs would be unstable with respect to any type of disorder [14, 13]. However recent theoretical work has suggested the opposite, namely that their surface conductivity is even stabilized by random disorder [15, 16, 17, 18, 19, 20, 21]. Here, the first ever synthesized WTI, namely $Bi_{14}Rh_3I_9$ [32] is used in order to characterize the distinct topological properties of this particular class of materials experimentally. Namely, STS measurements performed at 6 K will show that WTI exhibits helical and, thus, back-scatter free edge states at each step edge of its cleavage plane. Since such edge states can be intentionally created by AFM, this opens up unique possibilities of a topological protected 1D quantum circuitry. Most of the data described in this chapter are published in ref. [33].

5.1 General description of the first synthesized weak topological insulator $Bi_{14}Rh_3I_9$

As already reported in section 2.2.6, one way to create a WTI is the stacking of 2D TIs with topologically protected edge states. The side faces of the 3D system, formed by the step edges of the consecutive 2D TI layers then exhibit anisotropic surface states. The top and the bottom surface, however, remain dark, i.e. no surface states are present in the topological gap (cf. sketch in Fig. 5.1 c)). A material which has been proposed to be a candidate for a WTI and which is built by a stacking of 2D intermetallic layers, is $Bi_{14}Rh_3I_9$ [32, 187]. The crystal structure of the compound as derived from XRD is displayed in Fig. 5.1 a). It shows a periodic alternation of two distinct layers, namely a graphene-like honeycomb lattice formed by rhodium centered bismuth cubes (red $[(Bi_4Rh)_3I]^{2+}$ layer in Fig. 5.1 a)) and an insulating $[Bi_2I_8]^{2-}$ layer built by I-Bi zigzag chains (blue layer in Fig. 5.1 a)). The latter is acting as an insulating spacer layer, which reduces the coupling between two consecutive graphene-like layers. The graphene analogues of the $[(Bi_4Rh)_3I]^{2+}$ layer becomes vis-

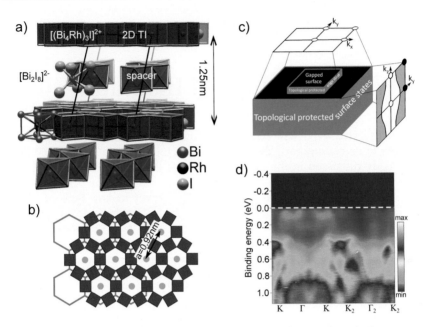

Figure 5.1: a) Triclinic atomic model as deduced from XRD revealing the layer structure of $Bi_{14}Rh_3I_9$. Insulating layers of $[Bi_2I_8]^{2-}$ (spacer layer) built by I-Bi zigzag chains separates the intermetallic $[(Bi_4Rh)_3I]^{2+}$ layers. b) Top view of the intermetallic layer which consists of Rh centered Bi cubes (red cubes) which form a honeycomb lattice and which is found to be a 2D TI by DFT [32]. The honeycomb lattice of graphene scaled by a factor ~ 3.8 is underlaid. c) Corresponding schematic of a WTI with the $\mathbb{Z}_2$ invariant 0;(001). The top surface of the WTI is gapped whereas the surrounding surfaces exhibit topological protected surface states. The surface Brillouin zones for the top and the side surface are marked, respectively, with the time-reversal polarization of the TRIMs marked as black (-) and white (+) dot and the resulting surface state (black line) separating green and white areas. The protected edge state at the rim of an island of the topologically trivial surface is also sketched. d) Energy dispersion of $Bi_{14}Rh_3I_9$ measured by ARPES, and revealing an energy gap at -170 to -370 meV below E_F (ARPES from ref. [32]). (Models from (a) and (b) derived by Bertold Rasche).

ible from Fig. 5.1 b), which shows a top view of the layer together with a conventional graphene lattice scaled by a factor of ~ 3.8 (gray structure). It becomes apparent that each honeycomb of the $[(Bi_4Rh)_3I]^{2+}$ layer is built by six Rh centered Bi cubes including a central iodide atom. The nodes of the graphene net are located in the centers of the prismatic voids originating from three conjoining Bi cubes. Unlike graphene, which has been suggested to be a 2D TI [42], but has not been confirmed experimentally due to the lack of strong SO coupling (band gap of only a few μeV [48]), the honeycomb lattice in $Bi_{14}Rh_3I_9$ is built by heavy elements, such that a strong SO coupling is present within the layer. Thus, we are dealing with a graphene-like band structure, however, in the presence of strong SO coupling, ingredients which have been proposed to enable a topological non-trivial quantum spin Hall state [42]. And indeed, a fully relativistic band structure calculation of the $[(Bi_4Rh)_3I]^{2+}$ layer reveals the presence of two gaps close to the Fermi level, which, by calculating their corresponding $\mathbb{Z}_2$ invariants, are found to be topological non-trivial[1] (Fig. 5.2 a)). Hence, from theoretical considerations, the honeycomb lattice of $Bi_{14}Rh_3I_9$ is a 2D TI. The calculations presented within this chapter are done by K. Koepernik and M. Richter from the group of Prof. J. van den Brink (IWF Dresden) and by B. Rasche from the group of Prof. M. Ruck (TU Dresden).

The topological nature of the whole compound $Bi_{14}Rh_3I_9$ has been analyzed in the work of Rasche et al. [32]. Owing to the weakly coupled stacking of the honeycomb lattice which is a 2D TI, $Bi_{14}Rh_3I_9$ is expected to be a WTI. Scalar and fully relativistic band structure calculations which are plotted in Fig. 5.2 d) and e) confirm this by showing the creation of a band gap of $\sim$ 210 meV due to SO interaction. The Dirac cone, which is present in the calculation without SO has been gapped out. Corresponding sketches of the band structure without and with SO interaction are shown in Fig. 5.2 b) and c). Again, the similarity to graphene when SO is effectively switched off becomes apparent. Note, that in the scalar relativistic calculation for $Bi_{14}Rh_3I_9$, the Dirac cone is located along ΓX and not at the K-point as in graphene. This results from the alternate stacking of the highly symmetric 2D TI layer and the low symmetric spacer layer which leads to a reduction of the overall crystal symmetry to the triclinic space group [32]. Moreover, Rasche et al. observed that the Dirac cones in the scalar relativistic calculation only show a minor dispersion perpendicular to the plane, rendering the calculated band structure quasi-2D and pointing to a very weak coupling of the stacked layers [32].

As in the case of a single $[(Bi_4Rh)_3I]^{2+}$ layer, the direct calculation of the four topological $\mathbb{Z}_2$ invariants $\nu_0;(\nu_1\nu_2\nu_3) = 0;(001)$ for the the energy gap in Fig. 5.2 e) classifies the material into the topological class of a WTI [32]. Interestingly, a second non-trivial energy gap at higher positive energies appears due to SO interaction as it is not visible in the calculation without SO. The $\mathbb{Z}_2$ indices of $\nu_0;(\nu_1\nu_2\nu_3) = 0;(001)$ imply that surfaces which are parallel to the 2D TI layers (perpendicular to the normal (001)), i.e. the natural cleavage

---

[1]Details on the computation present in this chapter are described in section 5.2.6.

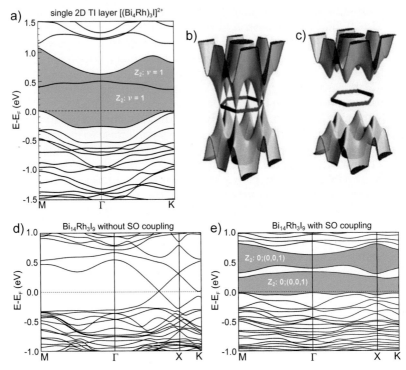

Figure 5.2: a) Fully relativistic DFT band structure of a single 2D TI layer $[(Bi_4Rh)_3I]^{2+}$. The green areas mark the topological band gaps with the calculated $\mathbb{Z}_2$ indices as marked (DFT by Bertold Rasche). b), c) Sketches of the Dirac cones in the situation without SO coupling (similar to the graphene band structure) (b) and the opening of the topological gap with SO coupling (c). d) DFT without SO coupling for $Bi_{14}Rh_3I_9$ with a Dirac cone present on the ΓX line. e) Fully relativistic band structure of $Bi_{14}Rh_3I_9$ where SO coupling opens up two topological band gaps with the calculated $\mathbb{Z}_2$ indices as marked. ((b)-(e) from ref. [32]).

planes in $Bi_{14}Rh_3I_9$, have no topological surface states and are the so-called dark surfaces [20], whereas at any other surface an even number of Dirac points appears (cf. sketch in Fig. 5.1 c)), each expecting to have a strongly anisotropic dispersion around it due to the quasi-2D nature of the bulk band structure.

Rasche *et al.* provided first photoemission data of the band structure of $Bi_{14}Rh_3I_9$ resolving a band gap at -170 to -370 meV below the Fermi level (Fig. 5.1 d)). Good agreement is found with the fully relativistic band structure calculation, consistent however with a highly n-doped nature of the material, which is ascribed to a slight deficiency of iodine at the surface. An overall energy shift of $\sim 400\,\text{meV}$ is found with respect to DFT [32]. The ARPES results in combination with the DFT calculations already provides some experimental hints for the presence of weak topological nature in $Bi_{14}Rh_3I_9$, however a direct identification of WTI properties is lacking. Therefore, ARPES measurements on surfaces which are perpendicular to the dark surface would be required. Such faces, however, do not belong to the natural cleavage plane of the crystal and are thus difficult to realize. On the contrary, a different approach for the experimental characterization of WTI properties is the detection of the surface states at the step edges of the natural cleavage plane (dark surface). They then appear as topological edge states directly at the step edge of the 2D TI layer [20] (cf. sketch in Fig. 5.1 c)) and are thus accessible by a local microscopic technique like STS.

5.2 Edge states at the dark side of the weak topological insulator $Bi_{14}Rh_3I_9$ probed by STS

This chapter starts with the identification of the different layers in $Bi_{14}Rh_3I_9$ by means of STM and STS. Each layer will be characterized by its atomic and electronic structure, and bias-dependent structural variations of the morphology which arises due to the coupling between the 2D TI layer and spacer layer will be discussed. Further, first experimental measurements of the topological edge states in a WTI as well as their unique properties which could be favorable for possible novel types of information processing will be provided by STS. Moreover, the tailoring of well defined step edges into the topologically dark surface of $Bi_{14}Rh_3I_9$ using AFM is demonstrated, showing that the material might be suitable for the construction of quantum networks exploiting the protected nature of the edge state. Further, I will show in this chapter, that an edge state is absent in the structural closely related, but topologically trivial insulator $Bi_{13}Pt_3I_7$ due to a pairing of adjacent layers, highlighting the topological nature of the edge state in $Bi_{14}Rh_3I_9$.

5.2.1 Identification of the atomic and electronic structure of the different layers in $Bi_{14}Rh_3I_9$ by STM

The $Bi_{14}Rh_3I_9$ crystals which are analyzed within this work has been prepared by Bertold Rasche from the group of Prof. Dr. M. Ruck from the Department of Chemistry in Dresden (Germany). The crystals were grown by high-temperature annealing of a stoichiometric mixture of Bi, Rh, and BiI3 (molar ratio 11: 3: 3). The starting materials were ground under argon atmosphere

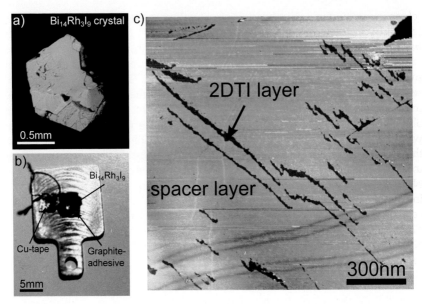

Figure 5.3: a) Optical microscope image of the $Bi_{14}Rh_3I_9$ crystal. Flat areas suitable for STM measurements are visible. b) $Bi_{14}Rh_3I_9$ crystal mounted to the molybdenum sample holder using a graphite-based adhesive. Cu-tape used for the cleavage in UHV is marked. Image has been taken after cleavage. c) STM overview image of the freshly cleaved $Bi_{14}Rh_3I_9$ surface ($V = 1\,V$, $I = 100\,pA$). Two different contrasts are visible which are attributed to the two distinct layers of the material as marked.

in a glovebox. The homogeneous powder was sealed in an evacuated silica ampule and heated to $700\,°C$ in a tubular furnace at a rate of approximately $600\,K/h$. Fast cooling to $420\,°C$ (cooling rate $-4\,K/min$) and then slow cooling to $365\,°C$ (cooling rate $-1\,K/h$) followed instantaneously. After three days, the ampule was quenched in water. This technique yields black, platelet-shaped crystals of $Bi_{14}Rh_3I_9$ with sizes up to $1 \times 1 \times 0.2\,mm^3$. An optical microscope image of the crystal is shown in Fig. 5.3 a) revealing the small size of the crystal. Obviously, large flat areas within the surface are visible which are suitable for the positioning of the tungsten tip in the STM experiment. Prior to the measurements, the $Bi_{14}Rh_3I_9$ crystal has been mounted to the sample holder using a graphite-based adhesive (Fig. 5.3 b)). Here, one had to take care of the right choice of adhesive as the iodide in $Bi_{14}Rh_3I_9$ compound is highly reactive, whereas a reaction with the adhesive could lead to a change of sample properties. As $Bi_{14}Rh_3I_9$ is a layered structure with a weak coupling between the layers, the same cleavage process was used as in the preparation of Sb_2Te_3

(cf. section 4.2.1). The Cu-tape was pushed on the Bi$_{14}$Rh$_3$I$_9$ surface prior to the transfer of the sample into the UHV chamber and later pulled down at a base pressure of $1 \cdot 10^{-10}$ mbar.

A first STM overview image of the freshly cleaved sample surface is depicted in Fig. 5.3 c). It shows the topologically dark (001) surface and reveals two different contrasts which are attributed to the two distinct layers of the material. As we will see below, both layers are clearly distinguishable in the STM due to their different atomic arrangement and different step heights, which makes the identification of each layer rather straightforward. Taken this into account, the upper layer (bright contrast) has been identified as the insulating spacer layer and the lower layer (dark contrast) as the 2D TI layer. Notice, that approximately 80 % of the top surface is covered by the insulating spacer layer which considerably complicates the search for the topological edge state which is predicted to be located at the step edge of the 2D TI layer. The surface coverage has been checked on several other Bi$_{14}$Rh$_3$I$_9$ crystals but always revealing the same unfavorable ratio of spacer layer and 2D TI layer. Moreover, the appearance has also been found to be independent of different cleavage approaches.

However, after some search, areas which show step edges of the 2D TI layer could be located. The typical appearance of step edges of both layers is visualized in the STM topography image of Fig. 5.4 a). A sequence of 4 consecutive layers is observed in the image whereas each layer can be identified by its specific step height (see inset of Fig. 5.4 a)). The STM step height of two adjacent layers combined exhibits ~ 1.2 nm and corresponds to the respective layer thickness deduced from XRD (1.25 nm) (cf. Fig. 5.1 a)). Thus, in comparison with step heights from the model, one is able to attribute the measured STM step height of ~ 0.4 nm to the insulating spacer layer and the height of ~ 0.8 nm to the 2D TI layer. A more direct identification of the respective layers, however, is realized by the atomic appearance of each layer. Figure 5.4 c) and d) show atomically resolved STM images recorded on both layers of the surface as marked by the respective colored box in Fig. 5.4 a). One layer (Fig. 5.4 c)) exhibits a graphene-like honeycomb lattice with a unit cell of ~ 0.9 nm. It is identified by overlaying the polyhedron model deduced from XRD and shows good agreement with the measurement. Obviously, single atoms of the Rh centered Bi cubes which form the honeycomb can not be resolved as a rather integrated signal throughout each honeycomb is found. This fact can be explained by the chemical bonding within the honeycomb lattice giving rise to delocalized states akin to the well known metallic states of benzene rings. The Bi-Rh bonds within the cubes are strongly localized, as are the bonds between the three-centered Bi atoms which form the bases of the triangular-prismatic voids (cf. Fig. 5.1 and section 5.1). All together, a quasi-2D bimetallic network with a rather constant electron density is formed

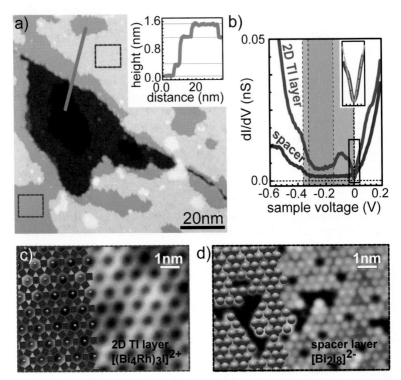

Figure 5.4: a) Atomically resolved STM image ($V = 1$ V, $I = 100$ pA) showing the layered structure of $Bi_{14}Rh_3I_9$. Inset: height profile along the green line revealing the different step heights of the layers. b) Typical $dI/dV(V)$ spectra for the different layers as labeled by the respective color ($V_{stab} = 0.8$ V, $I_{stab} = 100$ pA, $V_{mod} = 4$ mV). The energy gaps for each layer found in the spectra are marked by shaded colored boxes. Inset: zoom into a narrow energy region around E_F (-30 to 30 meV) for the 2D TI spectrum. Both sides increase linearly (marked by the linear fit in green). c) and d) Zoom into the marked areas of (a) displaying the two different layers. Atomic model structure is overlaid providing good agreement with the measurement (color code as in the model in Fig. 5.1 a)). The 2D TI layer (c) and the insulating spacer layer (d) are identified. ((c) $V = 1.5$ V, $I = 100$ pA, (d) $V = -1.3$ V, $I = 100$ pA).

throughout the honeycomb lattice leading to the observed appearance in the STM image [32]. Later I will come back to the atomic structure of the honeycomb lattice and reveal that the underlying spacer layer considerably affects the appearance of the 2D TI layer in the STM at different voltages.

The other layer (Fig. 5.4 d)) shows hexagonally arranged spots, which are identified as the iodide atoms from the spacer layer by comparison with the superimposed model. The I-I atomic distance of $\sim 0.45\,\text{nm}$ found in the STM experiment nicely fits with the atomic distance deduced from the XRD ($\sim 0.45\,\text{nm}$). Thus, each layer of the $Bi_{14}Rh_3I_9$ crystal is identifiable by its atomic structure.

Beside the atomic appearance, both layers also differ in their respective electronic structure. Figure 5.4 b) shows typical $dI/dV(V)$ spectra for the 2D TI layer (red curve) and the spacer layer (blue curve). As expected for an insulating layer, a larger gap is found for the spacer ($\Delta V \approx 350\,\text{mV}$) whereas the 2D TI layer exhibit a gap between $V = -180\,\text{mV}$ and $V = -360\,\text{mV}$ which is in excellent agreement with the gap found in the ARPES measurement (Fig. 5.1 d)). Moreover, the $dI/dV(V)$ spectrum of the 2D TI layer exhibits vanishing dI/dV intensity at E_F surrounded by a linear increase of the LDOS on both side (cf. inset in Fig. 5.4 b)). This type of gap is attributed to a 2D Coulomb pseudo-gap of Efros-Shklovskii type originating from electron-electron Coulomb interaction within a localized 2D system which reduces the

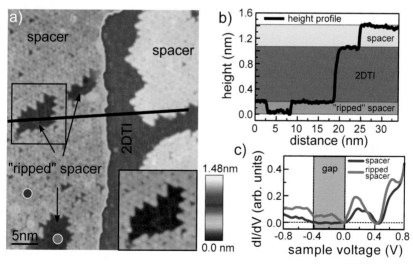

Figure 5.5: a) STM overview image of the different layers as marked ($V = 1\,\text{V}$, $I = 100\,\text{pA}$). Inset: zoom into the area marked by a black box. Both layers show the atomic arrangement of the spacer layer. b) Height profile of the layers marked by the green line in (a). A ditch within the spacer layer is identified by its reduced step height and marked as "ripped" spacer in (a) and (b). c) $dI/dV(V)$ spectra ($V_{stab} = 1\,\text{V}$, $I_{stab} = 100\,\text{pA}$, $V_{mod} = 4\,\text{mV}$) taken at the spacer layer (blue curve) and at the "ripped" spacer (turquoise curve) with the band gap of the spacer layer marked by the shaded box. The exact position of the $dI/dV(V)$ measurements are labeled in (a) by the respective color.

density of states near E_F [188]. This feature is a further sign for the quasi-2D nature of the honeycomb lattice, hence, the small interlayer coupling.

Identification of half a spacer layer

Fig. 5.5 a) shows an overview image of a stack of two spacer and one 2D TI layer. Interestingly, the lower spacer layer seems to be disrupted at some areas (marked by arrows). However, from the atomic corrugation, one observes that the layer underneath is not a 2D TI layer, but still shows the atomic arrangement of the spacer layer (inset in Fig. 5.5 a)). The corresponding height profile which is marked by the dark line and displayed in Fig. 5.5 b) reveals a step height of $\sim 0.2\,$nm for the area where the spacer layer is disrupted. This number corresponds to half an ordinary spacer layer step height. So obviously, the Bi-I bonding within the spacer layer is broken by the cleavage process leaving half of the spacer apart. Moreover, if we have a look at the electronic structure of the "ripped" spacer in comparison with the ordinary spacer by STS (Fig. 5.5 c)), one can see that the "ripped" spacer shows a finite dI/dV intensity in the energy range where the ordinary spacer exhibits a band gap. The course of the remaining $dI/dV(V)$ curve however shows general agreement. Consequently, the insulating character of the layer is obviously only reached if the spacer is complete.

5.2.2 Bias-dependent structural variations of the morphology of the 2D TI and spacer layer in $Bi_{14}Rh_3I_9$

In the following section, I will give a closer look at the atomic structure of both the 2D TI and the spacer layer. In order to give an accurate interpretation of the measured STM topography images, one has to keep in mind that the tunneling current is mostly proportional to the density of states of the sample (cf. eq. 3.3). Thus, the local density of states of the layers at specific energies has a huge effect on the appearance of the atomic lattice in the STM image.

Atomic contrast of the insulating spacer layer

Figure 5.6 shows a series of STM topography images of the spacer layer measured at the same area and mapped at different bias voltages. As it is obvious from first sight, the atomic appearance of the layer considerably depends on the applied bias voltage between the tip and the sample. The applied voltage plays an important role in the appearance of the image as it determines the range of the densities of states contributing to the image. Thus, regarding equation 3.3, all the states from E_F up to the applied voltage contribute to the tunneling current I and thus to the STM image. Hence, one needs to get some

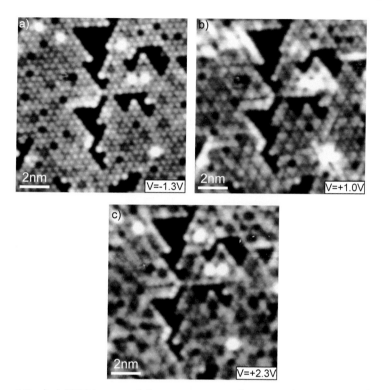

Figure 5.6: a)-c) STM images of the same area on the spacer layer measured with varying sample voltage, (a) ($V = -1.3$ V, $I = 200$ pA), (b) ($V = +1$ V, $I = 200$ pA) and (c) ($V = +2.3$ V, $I = 200$ pA). Bias-dependent appearance of the spacer layer lattice is visible.

information about local density of states of the layers at the particular bias voltages in order to interpret the different appearances in the STM images. Figure 5.7 shows fully relativistic DOS calculations in an energy range from - 3 to 3 eV for the individual layer atoms, for a) the 2D TI layer and b) the spacer layer. The black curve in each plot displays the average DOS for the whole compound $Bi_{14}Rh_3I_9$. Obviously, the Bi 6p states of the 2D TI layer (red curve in Fig. 5.7 a)) contribute the most to the average density at the unoccupied states, whereas the I 5p states of the spacer layer (light blue curve in Fig. 5.7 b)) provide the dominant part of the overall density for the occupied states. Interestingly for the spacer layer, the density of the I 5p states at the occupied states considerably exceeds the part of the Bi 6p states (orange curve in Fig. 5.7 b)), whereas at the unoccupied states, the contribution of both elements

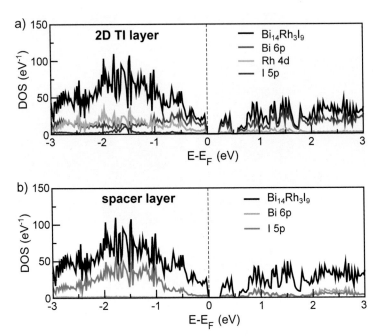

Figure 5.7: a) Fully relativistic DFT calculation of the projected DOS to specific elemental orbitals for the individual elements of the 2D TI layer. b) The same for the spacer layer. The symmetry of the contributing orbital for each element is marked for (a) and (b). (DOS calculation by Bertold Rasche).

is equal, indicating a strong hybridization between these two orbitals. This hybridization leads to more delocalized wave-functions in line with the more blurred appearance of the atomic structure at positive sample voltage (cf. Fig. 5.6 b) and c)). Further, a larger gap is visible just above E_F, whereas a second, smaller gap is located at $E - E_F = 0.5$ eV. With respect to the $dI/dV(V)$ spectrum of the 2D TI layer in Fig. 5.4 b) the calculation is shifted by ~ 350 meV which is inline with the findings in the work of Rasche *et al.* [32]. Moreover, the small maximum in the DOS between the two gaps in Fig. 5.7 arises from the Bi 6p states of the 2D TI layer and is absent in the DOS of the orbitals from the spacer layer. This fits again with the measured $dI/dV(V)$ spectra (Fig. 5.4 b)) which show an increase in the dI/dV signal above the gap and below E_F only for the 2D TI layer, whereas on the spacer layer the $dI/dV(V)$ intensity stays low over the whole energy range from -0.4 eV to E_F.

With the above DOS calculation for the spacer layer, one may now interpret the STM images measured at different bias voltages from Fig. 5.6. Consequently, the atoms visible in Fig. 5.6 a) of the spacer layer, which are measured at a bias voltage of $V = -1.3\,V$, can be assigned to the iodide atoms if compared to the calculation in Fig. 5.7 b). However, no preferred triangular shaped alignment of the iodide atoms as expected from the model in Fig. 5.1 a) is visible, probably due to the lack of Bi 6p intensity, preventing the visibility of the octahedral coordination within a Bi-I zigzag chain. At positive bias voltages (Fig. 5.6 b) and c)), the appearance of the spacer layer alters. At $V = +1\,V$ (Fig. 5.6 b)) where, according to the calculation, both elements I and Bi contribute equally to the DOS, the STM image shows a pattern where the triangular appearance of the Bi bonded trimers of I is directly visible. Thus, one get access not only to the iodide atoms on top, but also to the position of the underlying Bi octahedra. At even higher positive bias voltage ($V = +2.3\,V$), the density of the Bi 6p states is slightly dominating the I 5p states, resulting in the appearance shown in Fig. 5.6 c). Apparently, the appearance of the iodide lattice in the STM image slightly blurs as the contribution from the underlying Bi atoms gets stronger which can be interpreted as an increased delocalization of the corresponding wave-functions. Note, however, that also the tip sample distance increases with increasing bias voltage, so that the sharpness of the STM image decreases additionally.

Atomic contrast of the 2D TI layer

For the 2D TI layer, the STM results also reveal some considerable bias-dependent deviation from a perfect honeycomb lattice as present in Fig. 5.4 c). The reason is, however, different as in the case of the spacer layer. Namely, beside the energy dependent density of states projected to the different elemental orbitals, the arrangement of the underlying spacer layer considerably affects the density of states of the 2D TI layer and, thus, the appearance of this layer in the STM image. The vertical distribution of the wave-functions is thereby changed and, thus, their contribution to the tunneling current probing the LDOS in vacuum above the sample. Figure 5.8 a) and b) shows an atomically resolved area of the 2D TI layer at a bias voltage of $V = -300\,mV$ and $V = +300\,mV$, respectively, i.e. at much lower voltage than in Fig. 5.4 c), where a perfect honeycomb lattice is visible for the 2D TI layer. The remaining spacer layer within the area serves as a marker for the assignment of the same atomic rows in both images. Obviously, both images show a deviation from a perfect honeycomb lattice as well defined row patterns are observed. Moreover, by comparing both images recorded at opposite bias voltage, one observes that the atomic row shifts by one unit cell in the direction perpendicular to the atomic rows and half a unit cell parallel to the rows. This can be most easiest seen by following the gray lines in both images which mark

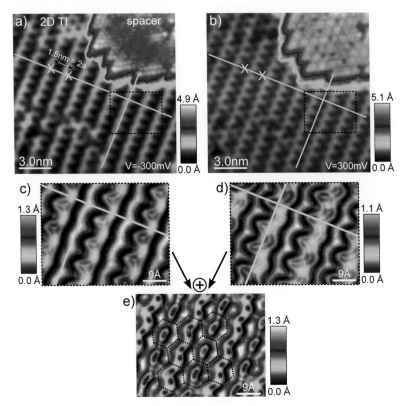

Figure 5.8: a), b) STM image of the 2D TI and spacer layer recorded at different bias voltages, (a) probing occupied states above the WTI gap and the area of the WTI gap (V = -300 mV, I = 80 pA) and (b) probing unoccupied states (V = +300 mV, I = 80 pA). The gray lines in both images mark the same positions. The distance between two atomic rows is marked. The atomic lattice of the spacer layer serves as a reference point for the gray lines. c), d) Zoom into the respective areas in (a) and (b) marked by the black dashed box. e) Summed image of the STM plots in (c) and (d) with a regular honeycomb lattice superimposed.

the same location[2]. Zooming into the area marked by a black dashed box in both images makes the shift of the lattice even more explicit (Fig. 5.8 c) and d)). In order to understand the row-like shape of the lattice as well as the bias-dependent shift by one unit cell, K. Koepernik simulated STM images by spatially resolving the densities of states restricted to specific energy intervals

---

[2]Thermal drift and piezo-creep is negligible in these data.

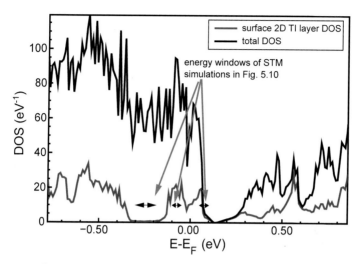

Figure 5.9: Fully relativistic DFT calculation of the projected DOS for a 4 layer slab terminated by a 2D TI layer (2D TI-spacer-2D TI-spacer). The red curve corresponds to the DOS of the surface 2D TI layer alone, and the black curve to the total density of states of the slab. For the surface 2D TI layer, the non-trivial gap is shifted by $\approx$ 300 meV into the occupied states with respect to the gap position in the bulk DFT calculation (Fig. 5.2 e)). The black double-arrows mark the energy regions of the simulated STM images shown in Fig. 5.10. (Calculation by Klaus Koepernik).

$\Delta E = [E_0, E_1]$ for a slab consisting of 4 layers and terminated by a 2D TI layer (2D TI-spacer-2D TI-spacer). The corresponding fully relativistic DOS calculation for the slab (Fig. 5.9) reveals a shift of the topological band gap by $\approx$ 300 meV into the occupied states for the top surface 2D TI layer (red curve) with respect to the bulk DFT calculation (Fig. 5.2 e)) and in agreement with the experimental observation of the cleaved surface (Fig. 5.4 b)), such that the gap is located at $\Delta E = [-300, -190]$ meV. However, the total DOS of the slab (black curve in Fig. 5.9) reveals a finite density of states within the energy region of the top surface 2D TI gap, which originates from the sub-surface spacer layer and the inner 2D TI layer. Their density of states is shifted in energy with respect to the surface 2D TI layer due to the surface electrostatic potential, such that their gaps do not coincide anymore.

The two top layers of the slab (2D TI layer with an underlying spacer) are shown in Fig. 5.10 d), where the black dotted box marks the area of the simulated STM images in Fig. 5.10 a)-c). The STM images correspond to the local density of states found for a plane being 1.94 Å above the last Bi atoms of the 2D TI layer and for an energy interval as labeled and as additionally marked

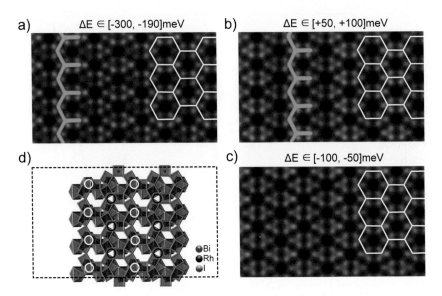

Figure 5.10: Spatially resolved density of states for the 2D TI layer of $Bi_{14}Rh_3I_9$ at a plane 1.94 Å above the last Bi layer for a) an energy region within the gap of the top surface 2D TI layer (ΔE = [-300, -190] meV), b) an energy region within the unoccupied states (ΔE = [+50, +100] meV) and c) an energy range below E_F but above the 2D TI layer gap (ΔE = [-100, -50] meV). For comparison, the regular honeycomb lattice is superimposed as white line pattern. In (a) and (b) contrast modulations within the honeycomb lattice are visible (superimposed straight line pattern is a guide to the eye) with a shift of one unit cell of the honeycomb lattice with respect to each other. The simulation in (a) has a much smaller absolute value of densities, which is not visible here as the images are normalized in contrast. d) Sketch of the topmost 2D TI layer (red) with the underlying spacer layer (blue). Dotted box marks the area of (a)-(c). The less bright triangles in (a) are those under which a blue layer octahedron lines up perfectly (positions marked by white circle), whereas in (b) the less bright triangles are those under which no contribution of the spacer layer is present (positions marked by black circle). (Calculations by Klaus Koepernik).

in the DOS calculation of Fig. 5.9 by black double-arrows. Notice that 1.94 Å is smaller than typical tip-surface distances implying better lateral resolution within the calculated data.

The bright dots in the simulated images are due to the Bi atoms. Obviously, small intensity modulations between the Bi atoms are visible for occupied states (Fig. 5.10 a)) and unoccupied states (Fig. 5.10 b)), as highlighted by a respective pattern which acts as a guide to the eye. The contrast indeed shifts about one unit cell in horizontal direction ($\sim$ 0.9 nm), and half a unit cell in vertical direction between the two plots. In contrast, no preferred orientation is observed for the energy interval in between the 2D TI gap and E_F (Fig. 5.10 c)). Here, the LDOS between the three-centered Bi atoms which form the bases of the triangular-prismatic voids are more intense than the Bi-Rh bonds within the Rh centered Bi cubes. Note that the image in a) has a smaller absolute value of densities because of being in the 2D TI gap. The pictures are however normalized in contrast, so that this information is not visible. Further, it was checked by DFT that this additional contrast shifting with bias is visible throughout the whole vacuum section up to 4.1 Å, where the signal gets too weak with respect to the noise within the numerical calculation.

The origin of the lattice modulation in the 2D TI layer can be explained by the orientation of the honeycomb lattice with respect to the underlying zigzag chains of the spacer layer. At energies corresponding to the band gap region of the top 2D TI layer, the total density of states decreases at locations where the Bi atoms of the triangular-prismatic voids lines up perfectly with a spacer layer octahedron (positions marked by white circles in Fig. 5.10 d)). On the other hand, at energies within the unoccupied states, the density obviously vanishes at positions where the Bi triangles have no immediate neighbor in the underlying spacer layer (positions marked by dark circles in Fig. 5.10 d)). The different orientations of the 2D TI layer with respect to the underlying spacer layer throughout the whole lattice thus leads to this periodic modulations in the considered energy intervals. A tentative explanation would be a hybridization of the Bi atoms of the 2D TI layer with the I atoms of the underlying octahedra, leading to occupied bonding configurations and unoccupied antibonding configurations. While the bonding configuration leads to a reduced density in vacuum, the antibonding configuration leads to an increased LDOS in vacuum with respect to the non-bonding configuration represented by the Bi atoms above the prismatic voids.

As the atomic modulation shifts by about exactly one unit cell of the honeycomb lattice between two images measured at opposite bias voltage, a mixing of the two plots in Fig. 5.8 c) and d), obtains an appearance which resembles a regular honeycomb lattice (Fig. 5.8 e)). Thus, the modulation of the honeycomb lattice, originating from the orientation between the 2D TI and the spacer layer is well defined and depends on the applied bias voltage in the STM experiment. Notice that the observed rows in the STM images are perpendicular to the rows of the Bi-I octahedra of the spacer layer allowing to trace the subsurface orientation by STM, as well as to easily determine the zigzag direction of the honeycomb layer. I checked about ten different locations on the Bi$_{14}$Rh$_3$I$_9$, as well as using different micro-tips, always revealing the same modulation shift between voltages lying within the occupied and unoccupied states. The appearance of an undisturbed honeycomb lattice for energies between the 2D TI gap and E_F (Fig. 5.10 c)) could, however, not been confirmed by the experiment. One reason is the lack of reliable data at such small bias voltage, as the probing tip was unstable due to the reduced distance between tip and sample. However, the few atomically resolved images in this particular bias range tend to exhibit a similar modulation as is present for higher negative voltages, thus not confirming the theoretical simulation of Fig. 5.10 c). STM images of the 2D TI layer with a regular honeycomb lattice as shown in Fig. 5.4 c) are rather rarely observed and also found to be independent of the applied bias voltage. It is likely that in this cases, the coupling between the 2D TI and the underlying spacer layer is possibly reduced due to a small detachment of the 2D TI layer by tip forces [189] or due to large defect areas within the spacer layer underneath the probed 2D TI layer.

5.2.3 Probing the edge state at the step edge of the 2D TI layer in $Bi_{14}Rh_3I_9$

In the following two sections, the focus will be on the detection and characterization of the topological edge state at the step edge of the 2D TI layer, which is the fingerprint of $Bi_{14}Rh_3I_9$ as a WTI distinguishing it from strong 3D TIs. This property is also required for the use of weak 3D TIs in 1D quantum networks. As already discussed above, the challenge is to find an appropri-

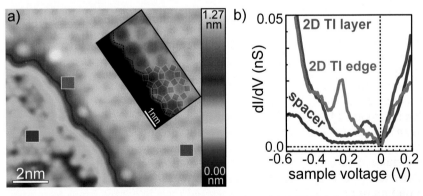

Figure 5.11: a) STM image with a typical step edge of the 2D TI layer ($V = 0.8$ V, $I = 100$ pA). Inset: zoom into the step edge region with overlaid honeycomb lattice from the model revealing the zigzag termination of the edge (dotted lines are guides to the eye). b) $dI/dV(V)$ spectra taken at the 2D TI (red curve), spacer layer (blue curve) and step edge of the 2D TI layer (gray curve). Positions of the recorded STS curves are marked in (a) by rectangles and labeled with the respective color ($V_{stab} = 0.8$ V, $I_{stab} = 100$ pA, $V_{mod} = 4$ mV).

ate step edge on the 2D TI layer as the sample surface is mostly covered by the insulating spacer layer (Fig. 5.3 c)). However, several 2D TI layer step edges could be detected on the course of the measurements and a typical one is depicted in Fig. 5.11 a). The 2D TI layer is again identified by its honeycomb lattice (yellow) and the spacer layer by the triangular-shaped appearance (blue). As in the case of most probed step edges, the edge exhibits a zigzag-termination (see inset in Fig. 5.11 a)) with a few kinks and a few adsorbates on top. I checked that the remaining adsorbates at the step edges are not originating from the background pressure in the UHV chamber by varying the background pressure during cleavage and transfer into the cryostat by a factor of ten. Furthermore, the time between the cleavage and the transfer into the cryostat was also varied by a factor of ten, with both not changing the amount of adsorbates at the step edges. Therefore, one can conclude that these adsorbates arise from the cleavage process itself and originate most probably from remainders of the spacer layer.

Figure 5.11 b) shows the $dI/dV(V)$ spectra taken on the different parts of the surface as marked by the different colored rectangles in Fig. 5.11 a). The electronic structure of the 2D TI and spacer layer have already been discussed in the previous sections and are again plotted as red an blue curve. Most importantly, the spectrum taken on the step edge of the 2D TI layer (gray curve) shows a strong dI/dV intensity within the energy gap of the 2D TI. The appearance of the respective $dI/dV(V)$ curves, i.e. the maximum in intensity for the curve on the step edge of the 2D TI layer and the minimum in intensity for the 2D TI and spacer layer, are typical for each region and are found on all areas of the sample, sometimes with a partly different intensity distribution, which is attributed to different local chemistry or to a different density of states of the probing tip.

Interestingly, the peak of the $dI/dV(V)$ curve measured at the step edge shows a maximum, which is slightly shifted towards lower energies with respect to the 2D TI band gap, thus, is further located at the bottom part of the band gap. Moreover, the detection of a topological state as a peak in the $dI/dV(V)$ signal differs from most STS measurements on strong TIs so far. The appearance of the topological state in the STS experiment depends on the energy dispersion of the state. In most strong TIs, the topological surface states form a Dirac cone with a minimum in density at the Dirac point. Thus, a minimum in the $dI/dV(V)$ measurement at the position of the Dirac point with a slight increase of intensity away from it is a fingerprint of a topological surface state in the STS experiment [80, 78, 73, 84]. In the case of Bi$_{14}$Rh$_3$I$_9$ however, the energy dispersion of the topological edge state looks different. Figure 5.13 d) shows a tight-binding calculation including SO interaction for a ruby lattice with a zigzag-terminated edge as it is for the 2D TI layer of Bi$_{14}$Rh$_3$I$_9$. At the bottom of the energy gap (lower energies), the edge state

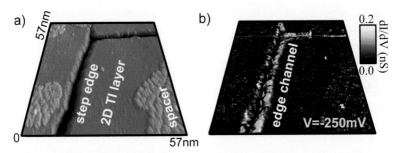

Figure 5.12: a) STM and b) spatially resolved dI/dV image measured at an energy within the 2D TI band gap and in a region containing step edges of the 2D TI layer as marked ($V_{stab} = -250\,mV$, $I_{stab} = 100\,pA$, $V_{mod} = 4\,mV$). Strong dI/dV intensity appears only at the step edges.

(colored in blue) exhibits no dispersion along Γ-K and only increases linearly towards the top of the gap. Thus, one would expect a maximum in the dI/dV intensity for the edge state at the bottom part of the 2D TI gap as indeed observed in the STS spectrum in Fig. 5.11 b). Note that the same tendency is observed for all other probed step edges.

The edge state, which has so far been detected by locally recorded $dI/dV(V)$ spectra, can further be analyzed using spatially resolved dI/dV maps (cf. section 3.1.3), which measures the local density of states of an area at a distinct energy. Figure 5.12 b) shows a spatially resolved dI/dV image at $V = -250\,mV$, i.e. at an energy within the 2D TI band gap. The corresponding topography image of the probed area is displayed in Fig. 5.12 a). It shows a large 2D TI layer area with a L-shaped trench crossing the layer. Bright stripes in the dI/dV image on both sides of the trench indicate the presence of an edge mode, as also found on all other probed step edges of the 2D TI layer. Note that this particular step edge does not run along a zigzag direction as can be easily seen from the stripes (see previous section) running from left to right in Fig. 5.12 b) which cut the step edge at an angle close to 90° and distinct from 60° and 120°.

In order to characterize some special properties of the edge state, a zoom of the step edge area of the 2D TI layer is shown in Fig. 5.13 a). It shows a spatially resolved dI/dV map at $V = -337\,mV$ providing a higher resolution image of the edge state (bright stripe). The corresponding topographic image of the 2D TI layer is shown in the inset. The Bloch type character of the state becomes apparent as an oscillation with unit cell periodicity (marked by olive rectangle in Fig. 5.13 a) and with corresponding profile height in c)) along the zigzag direction. Further, a very narrow spatial distribution of the edge state being confined to the last atomic row at the edge is observed. A profile

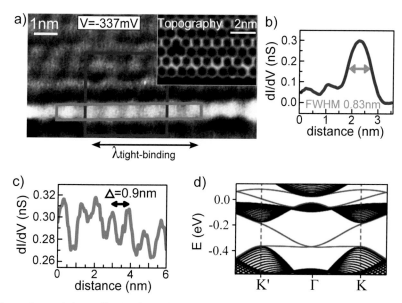

Figure 5.13: a) Spatially resolved dI/dV image within the bulk band gap (V_{stab} = -337 mV, I_{stab} = 100 pA, V_{mod} = 4 mV) of the step edge area of the 2D TI layer shown in the inset as a STM topography image (V = 0.8 V, I = 100 pA). Rectangles mark the areas of profile lines in (b) and (c). The double arrow marks the electron wave length of the edge state at this particular energy as deduced from the tight-binding calculation [190] in (d). b) Profile line perpendicular to the step edge and averaged in the parallel direction over the blue rectangle in (a) with FWHM of the edge state marked. c) Profile line along the step edge taken from the olive rectangle with marked peak distance corresponding to the size of one unit cell. d) Tight-binding calculation for a ruby zigzag edge terminated ribbon which is on the base of the 2D TI layer structure of $Bi_{14}Rh_3I_9$. The blue line marks the dispersion of the edge state in k-space. The energy scale has been adapted from the experimental data. (Tight-binding calculation from ref. [190]).

line across the edge state (blue square) exhibits a full width at half maximum (FWHM) of 0.83 nm only (Fig. 5.13 b)). This is an upper limit due to possible convolution effects with the tip shape. Thus, the edge state is confined to a single unit cell (a = 0.92 nm) as predicted by tight-binding calculations [190]. Such a width is much smaller than for edge states of the buried 2D TI made of HgTe quantum wells (edge state width: $\sim$ 200 nm) [7, 191] implying the possibility of much smaller devices.

The strong confinement of the edge state perpendicular to the edge is further illustrated in Fig. 5.14 a). The STM image shows a 2D TI step edge across

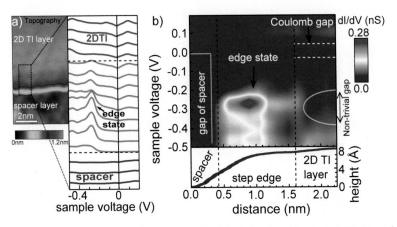

Figure 5.14: a) STM topography image ($V = 0.8\,\mathrm{V}$, $I = 80\,\mathrm{pA}$) of a 2D TI step edge. Inset: $dI/dV(V)$ spectra ($V_{\mathrm{stab}} = 0.8\,\mathrm{V}$, $I_{\mathrm{stab}} = 80\,\mathrm{pA}$, $V_{\mathrm{mod}} = 4\,\mathrm{mV}$) recorded at equal distances across the step edge within the marked dashed rectangle showing the evolution of the labeled edge state perpendicular to the step edge. b) Color plot of the $dI/dV(V)$ spectra from (a). Three different lateral regions are separated by dotted lines and labeled within the topographic profile below. Different energetic features are marked.

which locally resolved $dI/dV(V)$ curves have been recorded at equivalent distances. The curves are displayed in one plot with their area of origin labeled by the respective color. Again, it shows that the edge state is strongly confined to the very last atomic row of the zigzag edge even exhibiting some intensity at the slope of the step. The same data is displayed in Fig. 5.14 b) as a color code representation of the energy dependent LDOS across the step edge. The three different lateral regions are separated by dotted lines and labeled within the topographic profile below. Further energetic features, like the topological gap and the 2D Coulomb gap of the 2D TI layer become apparent as marked. One observes that pronounced edge state intensity is visible in the whole topological band gap and slightly weaker intensity even in the energy region below. This is in agreement with the tight-binding calculation (Fig. 5.13 d)), which shows that the edge state survives below the non-trivial gap.

5.2.4 Evidence for the topological nature of the edge state in $Bi_{14}Rh_3I_9$

So far, we have observed an edge state at the step edge of the 2D TI layer which is energetically located within the non-trivial band gap of the 2D TI

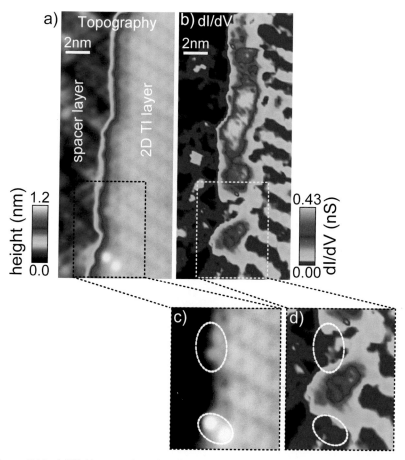

Figure 5.15: a) STM image of a rather disordered step edge of the 2D TI layer with the different regions as marked ($V = 0.8\,V$, $I = 80\,pA$). b) dI/dV image within the band gap of the 2D TI layer (averaged from $V = -180\,mV$ to $V = -350\,mV$, $I_{stab} = 80\,pA$, $V_{mod} = 4\,mV$) of the same area as (a). c), d) Zoom into (a), (b) as marked by rectangles. Dashed ellipses highlight the positions of the kink and the two adsorbates.

layer. Moreover, signs of an edge state at all step edges probed by STS was found with the edge state intensity present throughout the whole band gap region. In the following, I want to check the requirements imposed for a topological nature of the edge state in more detail. Namely, the topological protection of the edge state implies that the state is continuous along all step edges and continuous throughout the whole non-trivial band gap. Moreover, back-scatter processes due to scattering on impurities are prohibited due to the helical nature of the edge state, as long as the impurities do not break time-reversal symmetry. These properties all together pledge the topological edge state to be an example of what is often called a "perfectly conducting channel" [192, 20].

First, I want to analyze the spatial continuity of the edge state along the step edge of the 2D TI layer. A crucial point here is, that signs of the edge state must also be found at positions where adsorbates and impurities are located as no scattering or interruption of the edge channel due to topological protection occur, in contrast to trivial edge states. This behavior has been checked at about ten different step edges with probed lengths from 6 nm to 40 nm. Figure 5.15 a) shows exemplarily an STM image of a rather disordered step edge, exhibiting several kink positions and two adsorbates (marked by dashed ellipses in the zoom of Fig. 5.15 c)). The corresponding dI/dV intensity for the same area and for energies average over the 2D TI band gap region is plotted in Fig. 5.15 b). The intensity in the course of the edge state is reduced at positions of the step edge, where either a kink or two adsorbates are located (see dashed ellipses in Fig. 5.15 c) and d)). However, dI/dV intensity is observed surrounding these locations towards the interior of the 2D TI layer. The displacement of the dI/dV intensity indicates that the edge channel is pushed away from the step edge and simply moves around the obstacle [15]. The decrease in intensity shows that the edge state is additionally broadened in all three directions.

Figure 5.16 shows the same dI/dV image as in Fig. 5.15 b) however, in a different color code, including locally resolved $dI/dV(V)$ spectra ((i)-(v)). These spectra are recorded on the critical locations of the step edge (as marked in the dI/dV image by the respective color) in order to resolve the exact evolution of the edge mode in these areas. The spectra taken on the 2D TI layer in region (i) are flat within the band gap region with slightly different intensities in different regions. The origin of the varying intensities becomes directly visible in the dI/dV image by the diagonal stripe-like intensity fluctuation within the interior of the 2D TI layer. As already discussed in section 5.2.2, the stripes originate from the different coupling of the honeycombs of the 2D TI to the underlying zigzag chain pattern of the insulating spacer layer giving rise to a commensurability induced superstructure forming stripes along the zigzag direction of the honeycomb layer. This leads to an increased background intensity in the $dI/dV(V)$ spectra taken at positions of the stripes (e.g. green curve in Fig. 5.16 (i)) for energies of the 2D TI band gap and of the occupied bands. Comparing these spectra with the spectra at most parts of the step edge, for instance region (iii), a strong peak is revealed indicating the edge state. The peak maximum is again at the lower part of the band gap as expected from the edge state dispersion deduced by tight-binding calcula-

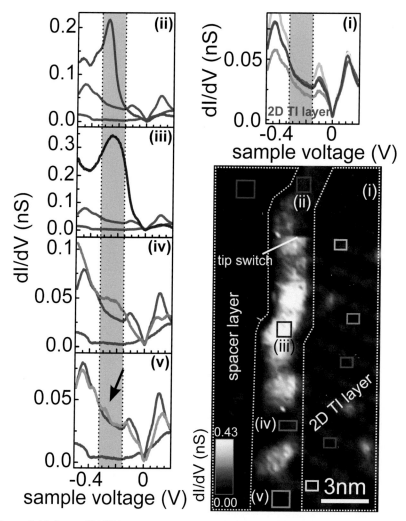

Figure 5.16: Same dI/dV image as in Fig. 5.15 b) displayed with a different color code including $dI/dV(V)$ spectra [(i)-(V)] ($V_{stab} = 0.8$ V, $I_{stab} = 80$ pA, $V_{mod} = 4$ mV) colored with respect to their area of origin marked in the dI/dV image. 2D TI layer and spacer layer are labeled and surrounded by dotted lines. The spectra in (i) originate from the 2D TI layer, the grey spectra in (ii)-(v) from the step edge region, the blue and red curves in (ii)-(v) from the spacer and the 2D TI layer, respectively. Shaded areas in (i)-(v) mark the bulk band gap as deduced from ARPES [32].

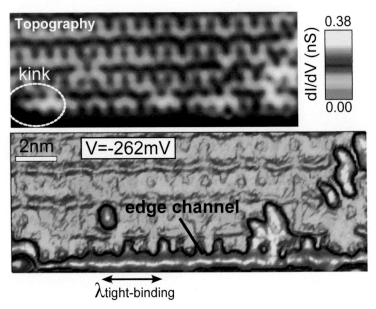

Figure 5.17: STM image (top, $V = -262\,\text{mV}$, $I = 100\,\text{pA}$) and corresponding dI/dV image at an energy within the band gap (bottom, $V_{\text{stab}} = -262\,\text{mV}$, $I_{\text{stab}} = 100\,\text{pA}$, $V_{\text{mod}} = 4\,\text{mV}$) of a rather straight step edge. A kink position is marked. The double arrow marks the expected electron wave length at this particular energy as deduced from tight-binding calculations (Fig. 5.13 d)) [190].

tions (cf. Fig. 5.13 d)). The sharpness of the peak depends on details of the tip, e.g., it appears sharper in region (ii) which is measured after a tip switch. Zooming into an area of small edge state intensity, region (iv), which is located around a kink position (Fig. 5.15 c)), reveals the same peak, but a factor of 10 lower in intensity with respect to region (iii), indicating a smaller coupling of the edge state to the tunneling tip. Moreover, the edge state intensity is pushed to the right, i.e. it moves around the obstacle as already observed in the other color code representation in Fig. 5.15 d) and as predicted by Ringel et al. [15]. The different geometry at some kinks may change the properties of the edge state, but does not affect its existence. Coming back to the area where two adsorbates are lying on the step edge (bottom part in Fig. 5.15 c)), rather no intensity was found in the dI/dV map. However, if one considers the $dI/dV(V)$ spectrum measured on top of the adsorbates (v), the size of the peak is further reduced, but can still be identified around $V = -270\,\text{mV}$ (see arrow), demonstrating that the edge state at least partly channels below these adsorbates. Thus, signatures of a spatially continuous edge state was found

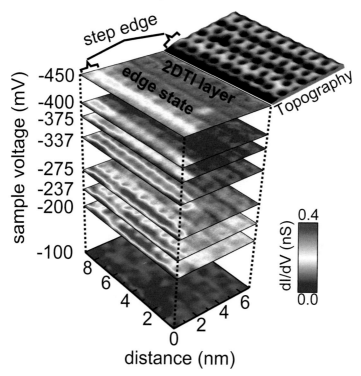

Figure 5.18: Stacked dI/dV images ($I_{\text{stab}} = 100\,\text{pA}$, $V_{\text{mod}} = 4\,\text{mV}$) of the area shown in the background STM image ($V = 0.8\,\text{V}$, $I = 100\,\text{pA}$) and recorded at voltages across the band gap as marked on the left.

within all investigated step edge areas, pointing to a robust character with respect to disorder as expected from topological protection.

Figure 5.17 shows another example of a 2D TI step edge (STM image on top) again with an extremely narrow edge state running along a straight edge and being pushed around a kink position (on the left of the image). Moreover, it shows that intensity fluctuations of the edge state along the edge are small as expected for the prohibited backscattering. In order to sustain this assumption, the electron wave length as deduced from tight-binding calculations (Fig. 5.13 d)) $\lambda_{\text{tight-binding}}$ is added to Fig. 5.17 and 5.13 a). Obviously, there is no structure with periodicity $\lambda_{\text{tight-binding}}/2$, i.e. no standing electron waves, in remarkable contrast to conventional 1D electron systems, where such oscillations exhibit LDOS intensity oscillations close to 100 % [193].

We have already observed in Fig. 5.14 b) that a pronounced edge state intensity is visible in the whole topological band gap. The energetic continuity is further visualized by the stack of dI/dV maps in Fig. 5.18. The corresponding STM image of the 2D TI layer is shown in the background. The edge state intensity is visible along the whole edge at all energies within the band gap and even present at energies below (-400 mV, -450 mV) however with weaker intensity. This covers the observation made on other step edges, as for example the edge state measured in Fig. 5.14 b), where slightly weaker intensity is also visible at energies below the band gap. However at energies above the energy gap, e.g. $V = -100\,\mathrm{mV}$ in Fig. 5.18, the edge state disappears.

Thus, the most important ingredients of a topologically protected edge state, continuity in space and energy, have been established. Further, one observes no indications of standing waves of the edge state electrons at the step edge, just as expected for topological edge states due to the prohibited backscattering. Notice that edge states barely prone to backscattering have also been observed on some of the step edges of the 2D TI Bi bilayers on Bi(111) [194] or Bi_2Te_3(0001) [195], but in both cases the edge states energetically overlap with bulk states, so that they intrinsically cannot be used for electronic applications. In contrast, the edge state of the weak 3D TI $Bi_{14}Rh_3I_9$ could provide perfect conduction channels at each step edge of the dark surface, if the Fermi level is brought to the topological non-trivial gap, e.g., by surface doping.

Scratching well-defined quantum networks into the surface of $Bi_{14}Rh_3I_9$ by AFM

Theory predicts that the helical and, thus, perfect conduction remains robust for step heights containing any odd number of exposed stacks, as also for even numbers stabilized by disorder [15, 16, 17, 18, 19]. Thus, simply scratching the surface deeper than a single layer tends to induce one-dimensional electron channels with a robust conductivity of at minimum e^2/h [17]. The topological nature of the edge state and, in particular, its prohibited backscattering will cause electrons with opposite spins moving in opposite directions at each of these step edges and being protected against localization. Eventually, tailored step edges could be connected to each other leading to quantum networks not suffering from backscattering and localization. To this end, the contact mode of the AFM is used to scratch partially straight step edges into the surface. The scratches are produced within a commercial AFM (Bruker) using a carbon coated silicon cantilever in AFM contact mode at ambient conditions (contact force during scratching: $F = 10^{-6}\,\mathrm{N}$). The AFM images have been recorded in the tapping mode using the same carbon coated silicon

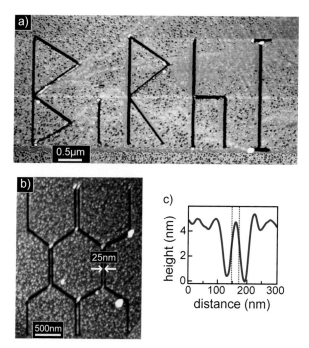

Figure 5.19: a) AFM image of $Bi_{14}Rh_3I_9$ surface with letters BiRhI scratched into the surface by a carbon coated silicon cantilever in AFM contact mode at ambient conditions (contact force during scratching: $F = 10^{-6}$ N). Average depth of the cuts ~ 15 nm. b) AFM image of the same surface after scratching a quantum network of step edges into the surface using the same parameters as in (a). c) Height profile of the area marked by the blue line in (b) reveals a distance of only ~ 25 nm between opposite edges. (Data by Bernhard Kaufmann).

cantilever (resonance frequency 275.1 kHz, force constant 43 N/m, oscillation amplitude 30 nm, set point 70 % and velocity 2 μm/s).

Figure 5.19 a) shows the chemical symbols of the material scratched into the surface, while Fig. 5.19 b) shows a network of AFM induced cuts which are about 3 layers deep and which are lead close to each other (cf. height profile in Fig. 5.19 c)). The distances between the centers of scratches are well below 100 nm and the edge channels are partly separated by 25 nm only. A further reduction of the distance between opposite step edges, short enough for electron tunneling, would lead to a beam splitter for electrons, which could come out of the splitter either left or right after traveling the parallel close-by step edges long enough. However, in order to make such networks operational, it is mandatory to move the non-trivial band gap to E_F. This may be achieved

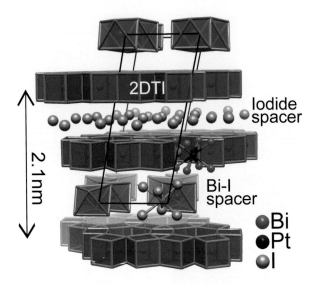

Figure 5.20: Atomic model of $Bi_{13}Pt_3I_7$ as deduced from XRD [196]. Two different spacer layers (pure iodide spacer and Bi-I spacer) are present, alternately spacing the 2D TI layers (red). The two different spacers lead to an alternating coupling between adjacent 2D TI layers, giving rise to a dimerization. (Model derived by Bertold Rasche).

by surface doping (e.g. by Iodine) recalling that the calculated bulk position of E_F is already within the non-trivial band gap (Fig. 5.2 e) and ref. [32]).

5.2.5 Absence of an edge state in the structural closely related but topologically trivial material $Bi_{13}Pt_3I_7$

In section 2.2.6, I discussed the topological protection of the surface state of WTIs and concluded that only a pairing of the 2D TI layers, in addition to an even number of layers may gap the surface states at the non-trivial surfaces of a WTI [17, 16, 15], if the pairing asymmetry is larger than the symmetry disorder. Whereas an even number of 2D TI layers alone, without any dimerization of the layers, is stabilized by any type of weak disorder of the interlayer coupling. By the STS data, the topological nature of an odd number of layers is confirmed by looking at step edges which are only 1 monolayer in height. In order to demonstrate the first assumption and additionally further consolidate the topological character of the edge states found in $Bi_{14}Rh_3I_9$, the very similar system $Bi_{13}Pt_3I_7$ is investigated, where a dimerization of the 2D

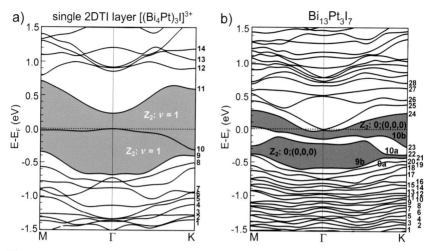

Figure 5.21: a) Fully relativistic DFT band structure of a single 2D TI layer $[(Bi_4Pt)_3I]^{3+}$ with numbered bands as used in Fig. 5.24 c). Green areas mark topological band gaps with calculated $\mathbb{Z}_2$ indices marked. b) Same as (a) for the 3D material $Bi_{13}Pt_3I_7$ using a modified spacer layer leading to the same lateral unit cell as for $[(Bi_4Pt)_3I]^{3+}$. Bands labeled as in the table of parities in Fig. 5.24 d). Pairs of bands are additionally labeled by the same numbers as in (a), but index a and b. Trivial band gaps corresponding to the non-trivial band gaps of the 2D TI are marked in orange with $\mathbb{Z}_2$ indices marked (The area between band 23 and 24 is not a real gap within DFT, since band 24 at Γ and band 23 at M overlap in energy). (DFT by Bertold Rasche).

TI layers takes place due to a replacement of Rh by the heavier Pt, resulting in the disappearance of the topological edge state.

Figure 5.20 shows the structural model of $Bi_{13}Pt_3I_7$ as deduced from XRD. The chemical composition is slightly different such that every second spacer layer is replaced by a single layer of iodide ions [196]. Thus, there are two distinct spacer layers, i.e. the structurally identical Bi-I spacer as in $Bi_{14}Rh_3I_9$ and a pure iodide spacer, leading to an alternating coupling between adjacent $[(Bi_4Pt)_3I]^{3+}$ honeycomb layers (red layer in Fig. 5.20). This honeycomb layer is again built by six Bi cubes, however with the central Rh replaced by the heavier Pt. Due to the second spacer layer, the size of one unit cell in stacking direction (2.1 nm) nearly doubles with respect to $Bi_{14}Rh_3I_9$.

The electronic structure of the honeycomb layer has been analyzed by fully relativistic DFT (Fig. 5.21 a)) and was found to be a 2D TI with two non-trivial band gaps (colored in green) as in the case of $Bi_{14}Rh_3I_9$. The three bands surrounding the non-trivial band gaps in the 2D TI layer of $Bi_{14}Rh_3I_9$ (Fig. 5.2

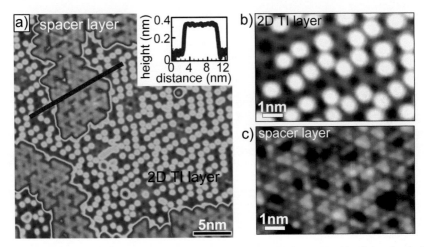

Figure 5.22: a) Atomically resolved STM image ($V = 1$ V, $I = 100$ pA) of the cleaved $Bi_{13}Pt_3I_7$ surface. Inset: Height profile along the green line. b), c) Atomically resolved STM images of the 2D TI and Bi-I spacer layers, respectively ((b) $V = 1$ V, $I = 100$ pA, (c) $V = 0.6$ V, $I = 100$ pA).

a)) look nearly identical to the bands 9, 10, and 11 of $[(Bi_4Pt)_3I]^{3+}$ except for a chemical shift by 0.3 eV upwards, which is caused by the different numbers of electrons within the 2D TI layers. Thus, the two materials probed in this study consist of very similar 2D TIs with honeycomb structure. The fully relativistic DFT band structure for the whole $Bi_{13}Pt_3I_7$ compound is displayed in Fig. 5.21 b). For the sake of simplicity, the arrangement of the Bi-I polyhedra within the spacer layer has been changed artificially, leaving the in-plane unit cell unchanged with respect to the original 2D TI layers. However, we checked that only small changes in the courses of the displayed bands are observed by performing non-relativistic DFT calculations (not shown here) of the real atomic arrangement as deduced from XRD. The band structure including the determination of the $\mathbb{Z}_2$ indices reveals the topologically trivial nature of $Bi_{13}Pt_3I_7$. Moreover, the material is semi-metallic as the two gap-like areas (colored in orange) are no real gaps in DFT since band 21 at the M-point and band 22 at the K-point for the lower gap, as well as band 24 at Γ and band 23 at M for the upper gap slightly overlap in energy. Thus, indeed, the alternating coupling between adjacent 2D TI layers renders the WTI topologically trivial as expected from analytic arguments and numerical toy models [15, 16, 17, 18, 19]. Later I will come back to the calculations, in order to analyze the precise mechanism which by dimerization renders $Bi_{13}Pt_3I_7$ trivial.

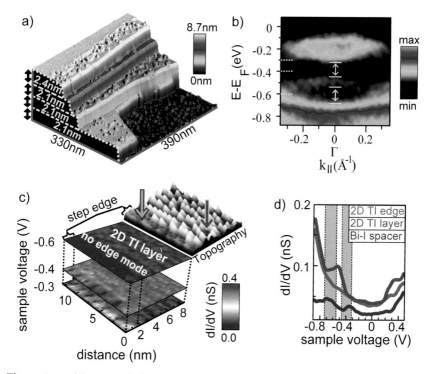

Figure 5.23: a) Large scale STM image ($V = 1$ V, $I = 100$ pA) showing step heights of one unit cell only (2.1 nm) as marked. b) ARPES intensity plot ($h\nu = 21.2$ eV); full lines with double arrows mark the two band gaps present, and the dotted lines the energies of dI/dV maps in (c). c) Stacked dI/dV images ($I_{stab} = 100$ pA, $V_{mod} = 8$ mV) of the area shown in the background STM image ($V = 0.6$ V, $I = 100$ pA) and recorded within the band gaps at voltages marked on the left, same contrast as in Fig. 5.18. d) Local $dI/dV(V)$ spectra ($V_{stab} = 1$ V, $I_{stab} = 100$ pA, $V_{mod} = 8$ mV) recorded at the positions marked by arrows in (c) and on the insulating spacer layer. Band gaps as deduced from ARPES are marked in red. (ARPES by Jens Kellner).

Similar to the experimental procedures on $Bi_{14}Rh_3I_9$, the surface of $Bi_{13}Pt_3I_7$ has been cleaved in UHV prior to the STM measurements. Figure 5.22 a) shows an overview image revealing two different layers. Again, the 2D TI layer is identified by its honeycomb structure (Fig. 5.22 b)) and the spacer layer by the hexagonally arranged spots, which are triangularly reconstructed (Fig. 5.22 c)). As the step height of the spacer (cf. inset of Fig. 5.22 a)) corresponds to the step height of the spacer layer in $Bi_{14}Rh_3I_9$, one can deduce that the spacer in Fig. 5.22 a) and c) represents the bigger Bi-I spacer. Further, one observes the same atomic appearance as for the identical $[Bi_2I_8]^{2-}$ spacer in $Bi_{14}Rh_3I_9$ which would not be expected for a pure iodide layer and which substantiates the above conclusion. Note, that the 2D TI layer exhibits single atoms on top, most probably remaining iodide ions from the spacers. Interestingly, such iodide ions are absent within the last two unit cells close to the zigzag-terminated step edges (Fig. 5.23 c), topography).

Fig. 5.23 a) shows a typical large scale STM image of the $Bi_{13}Pt_3I_7$ surface exhibiting several terraces interrupted by corresponding step heights. Most importantly, these step heights are always found to cover a complete unit cell in stacking direction (2.1 nm). The STM data also reveal an additional bigger $[Bi_2I_8]^{2-}$ spacer on the terraces (see topmost terrace with a step height of 2.4 nm), whereas a pure iodide spacer has not been observed throughout the measurements. Obviously, the coupling is stronger for adjacent 2D TI layers spaced by a pure iodide layer so that the cleavage takes place at the $[Bi_2I_8]^{2-}$ spacer. Consequently, one ends up with a unit cell of dimerized layers always resulting in an even number of 2D TI layers at each step edge.

Prior to the STS measurements, the experimental energy position of the pseudo band gaps found in the DFT were primarily determined using ARPES (Fig. 5.23 b)). The determination of these gaps is crucial since a possible absence or presence of edge state is expected within these areas. Remember that these gaps are found to be trivial in DFT calculations albeit originating from non-trivial gaps of the 2D TI (Fig. 5.21). The ARPES spectra were measured on the cleaved $Bi_{13}Pt_3I_7$ surface at 15 K using He I ($hv = 21.2\,eV$) discharge within a laboratory based UHV system. The overall energy resolution was 10 meV and the angular resolution 0.6°. The fact that the beam spot of the incident light is of the order of the sample size ($\sim$ 1 mm in diameter) causes some background intensity in the ARPES data probably originating from the carbon conductive adhesive. For the same reason, a slight softening of the bands in the spectra is observable. However, two band openings around the Γ-point are visible (marked by double arrows). With respect to the DFT calculations, the Fermi level is shifted down by about 0.3 eV indicating surface charging due to the cleavage process. The slightly smaller value compared to $Bi_{14}Rh_3I_9$ ($\sim$ 0.4 eV) might be explained by the remaining iodide ions on the surface of the 2D TI layer in $Bi_{13}Pt_3I_7$ (Fig. 5.22 a), b) and Fig. 5.23 c), topography), which reduces the n-type doping with respect to $Bi_{14}Rh_3I_9$.

Within the energy gaps found in ARPES, STS does not show any major sign of an edge state within all ten step edges probed. Figure 5.23 c) exemplarily shows dI/dV maps recorded at the 2D TI step edge at energies within the band gaps and as marked by dotted lines in the ARPES spectrum. Obviously, no edge state is present as also checked at other energies. Notice that the contrast in Fig. 5.23 c) is chosen identical to Fig. 5.18, where the edge state of

$Bi_{14}Rh_3I_9$ is clearly apparent. Identically, locally resolved $dI/dV(V)$ spectra at relevant positions show no edge state behavior within an energy range from -0.8 eV to 0.5 eV. Partly, the intensity of the 2D TI layer is even higher within the incomplete band gaps than the intensity found at the step edge. Thus, indeed, the dimerized structure of $Bi_{13}Pt_3I_7$, where stacks are built from pairs of 2D TIs, is a trivial insulator without protected edge states as predicted by topological analysis [15, 16, 17, 18, 19].

Topological analysis of $Bi_{13}Pt_3I_7$ by DFT

Here, I come back to the DFT results and the topological analysis of the trivial compound $Bi_{13}Pt_3I_7$ and its non-trivial $[(Bi_4Pt)_3I]^{3+}$ 2D honeycomb lattice. For both structures, the parities for each band at the corresponding TRIMs of the respective Brillouin zone have been determined including the corresponding $\mathbb{Z}_2$ indices valid for the energy region above the corresponding band (tabulated in Fig. 5.24 c) and d) for the $[(Bi_4Pt)_3I]^{3+}$ 2D TI layer and the whole $Bi_{13}Pt_3I_7$ compound, respectively). $\mathbb{Z}_2$ indices corresponding to band gaps are highlighted by dashed boxes. The band numbers are labeled according to the assignment of bands in the DFT band structure calculations (Fig. 5.21 a) and b)) and the TRIMs are marked in the representation of the respective Brillouin zone (Fig. 5.24 a) and b)).

In the case of the 2D TI layer, the first column of parities describes the Γ point, which is always a TRIM, while the other three points describe the remaining three TRIMs, i.e., the three different M points (see Fig. 5.24 a)). One observes that $\mathbb{Z}_2$ gets non-trivial at band 5 below E_F and gets trivial again at band 11 such that the marked band gaps around E_F are topologically non-trivial containing edge states. Consequently, the $[(Bi_4Pt)_3I]^{3+}$ layer is a 2D TI according to LDA-PW92 (see section 5.2.6).

The exchange of parities between bands is quite complex. Multiple avoided crossings between the different valence bands, all dominated by Bi 6p orbitals from the 2D TI layer, can be conjectured from the individual band courses. We have cross-checked that these avoided crossings also cause a change of orbital character (p_x, p_y, p_z) of the bands. A similar complexity in parity exchange has been found for the single 2D TI $[(Bi_4Rh)_3I]^{2+}$ layers in $Bi_{14}Rh_3I_9$ (not shown here), thus both 2D TI layers are very similar.

If one compares the DFT band structures of the $[(Bi_4Pt)_3I]^{3+}$ 2D TI layer and the whole $Bi_{13}Pt_3I_7$ compound in more detail (Fig. 5.21 a) and b)) a doubling of the bands, at least for bands 9 and 10 of the 2D TI layer become apparent. For this particular bands, the same band numbers were used in the respective plots with additional labels a and b in Fig. 5.21 b). The band doubling originates from the doubling of the unit cell due to dimerization. The pairs of bands exhibit exactly inverted parities at each TRIM indicating that

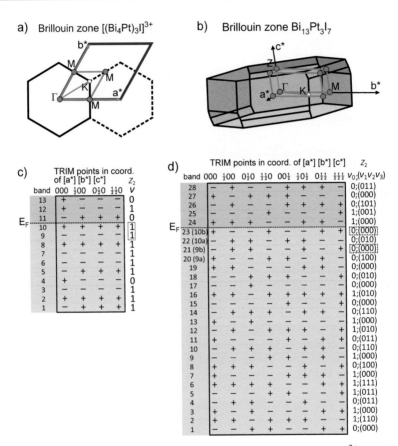

Figure 5.24: a) Brillouin zone (black hexagons) of the 2D TI $[(Bi_4Pt)_3I]^{3+}$ with high symmetry points (Γ, K, M) marked. The orange diamond connects the TRIMs (marked by blue points) and the blue lines show the corresponding unit cell. The k-space directions of the corresponding band structure displayed in Fig. 5.21 a) are marked by green lines. Coordinate directions a* and b* in reciprocal space are additionally marked. b) Same as (a) but for the 3D material $Bi_{13}Pt_3I_7$. The k-space directions of the corresponding band structure displayed in Fig. 5.21 b) are marked by green lines. c) Table of parities of the different bands as numbered in the band structure in Fig. 5.21 a) at the different TRIMs for $[(Bi_4Pt)_3I]^{3+}$. The $\mathbb{Z}_2$ indices valid in the energy region above the corresponding band are shown on the right. $\mathbb{Z}_2$ indices of band gaps are highlighted by dashed boxes. Directly below band 1, the topology is trivial. d) Same as (c) for the 3D material $Bi_{13}Pt_3I_7$. (Calculation of the band parities by Klaus Koepernik).

they represent the bonding and anti-bonding linear combinations of the corresponding original bands of the two 2D TIs which form the unit cell. Since the product of the parities at the TRIMs for each pair of such bands results in a minus sign, one gets a trivial band topology for each pair of bands. Thus, we suggest that a simple doubling of the unit cell renders the WTI trivial which is different to the mechanism proposed from most of the analytic theories, which consider the fate of the edge states only. Thus, one can conclude that a dimerization of adjacent layers does not only lead to a gap in the protected edge state, but at the same time destroys its protection by the symmetric and antisymmetric hybridization of the dimerized layers. This appears logical, since a destruction of a protected state requires a destruction of its protection, but should be rigorously analyzed in terms of topology separately.

5.2.6 Computational details

All band structure calculations shown in this chapter have been performed by our colleagues from the department of chemistry of the TU Dresden (Prof. Dr. Michael Ruck) and from the Leibniz Institute for Solid State and Materials Research of the IFW Dresden (Prof. Dr. Jeroen van den Brink). They were

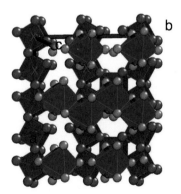

Figure 5.25: Sketch of a small part of the Bi$_{14}$Rh$_3$I$_9$ model visualizing the presence of inversion symmetry within the structure. The Rh centered Bi cubes of the 2D TI layer are colored in purple and the I-Bi zigzag chains of the spacer layer in blue. The red crosses on top of the Bi cubes mark the respective inversion centers. The crystallographic directions are additionally labeled by a, b and c.

performed using the Full-Potential Local-Orbital (FPLO) code [197] version 14.00, within the local density approximation (LDA) using the parametrization PW92 according to Perdew *et al.* [198]. The Blöchl corrected linear tetrahedron method with a 8×8×4 k-mesh for $Bi_{13}Pt_3I_7$ and a 12x12x1 k-mesh for the single 2D TI layer $[(Bi_4Pt)_3I]^{3+}$ was employed. For the 2D TI layer $[(Bi_4Rh)_3I]^{2+}$ of the $Bi_{14}Rh_3I_9$ compound, a 12x7x7 k-mesh has been employed. SO coupling is treated on the level of the four-component Dirac-equation. The following basis states are treated as valence states: Bi: 5s, 5p, 5d, 6s, 7s, 6p, 7p, 6d; Pt: 5s, 5p, 5d, 6s, 6p, 6d, 7s; I: 4s, 4p, 4d, 5s, 6s, 5p, 6p, 5d. For the band structure calculation of $Bi_{13}Pt_3I_7$, the atomic structure as deduced from XRD experiments was used [196]. The single 2D TI layer $[(Bi_4Pt)_3I]^{3+}$ was calculated with an iodide layer placed above and beneath the $[(Bi_4Pt)_3I]^{3+}$ layer in order to adjust the charge. 10 Å of vacuum was added in c-direction in order to separate adjacent layers in a 3D supercell geometry. The cell constants and atomic positions were optimized within the p6/mmm layer symmetry [32].

The calculation of the topological invariants was implemented following ref. [14] using the fact that the crystals all have an inversion symmetry (Fig. 5.25) such that parities of the states at the TRIMs can be used to calculate the four $\mathbb{Z}_2$ indices $\nu_0;(\nu_1\nu_2\nu_3)$ (cf. section 2.2.5 and references [13, 14, 49, 54]). Additional details can be found in the Supplementary Information of Rasche *et al.* [32].

6 Summary

Within this work, scanning tunneling spectroscopy and angle-resolved photoemission spectroscopy has been used in order to analyze different types of topology in relevant systems. The strong topological properties, which are already characterized for a broad range of materials, have been studied for the technologically relevant phase change materials. In the case of the weak topology, i.e. the other topological class in 3D, the experimental study on systems which are build by graphene-like sheets, displays the first ever demonstration of weak topological properties.

Using spin-ARPES at the synchrotron BESSY in Berlin, the spin texture of the Dirac cone within the fundamental gap of the phase change alloy Sb_2Te_3 has been observed, revealing a counter-clockwise rotation for the lower part of the Dirac cone. A spin polarization of up to 90 % could be detected after subtraction of the bulk valence band contribution in accordance with DFT calculation. Further, a Fermi velocity of $v_F = 3.8 \pm 0.2 \cdot 10^5$ m/s has been deduced, which agrees reasonably with $v_F = 3.2 \cdot 10^5$ m/s obtained by DFT. This result is also in line with the Fermi velocity deduced from the Landau level spectroscopy by STS, which provides a Fermi velocity of $v_F = 4.44 \pm 0.07 \cdot 10^5$ m/s for the topological surface state. In addition, the linear dependence of the Landau level energies with the root of the applied magnetic field $\sqrt{B}$ confirmed the Dirac fermion nature of the topological surface states as well as the field independent $n = 0$ Landau level of the Dirac point. Moreover, in accordance with DFT calculations, ARPES data further identified a novel, strongly spin-split Rashba-type surface state which is protected by a SO gap away from $\bar{\Gamma}$ and connects an upper and a lower bulk valence band. This state is similarly to the TI state protected by symmetry according to a fundamental criterion given by Pendry and Gurman in 1975.

A second phase change compound from the pseudobinary line of phase change materials, i.e. $Ge_2Sb_2Te_5$, which is at the borderline of the systems predicted to exhibit topological properties, has been analyzed by ARPES and STM. The experimental results reveal that metastable cubic $Ge_2Sb_2Te_5$ epitaxially grown on Si(111) exhibits a valence band maxima at $0.14 - 0.18$ Å^{-1} away from $\bar{\Gamma}$ and a band gap of 0.4 eV. All DFT calculations of $Ge_2Sb_2Te_5$ find a valence band maxima away from Γ only for a $\mathbb{Z}_2$ topological invariant $v_0 = 1$, which implies that the technologically most relevant phase change material

$Ge_2Sb_2Te_5$ is a strong topological insulator. This finding opens up the possibility of switching between an insulating amorphous and a topological phase on ns time scales.

In order to give a direct proof of the topological properties of $Ge_2Sb_2Te_5$ spin-resolved ARPES as in the case of Sb_2Te_3 is required. Moreover a better resolution of the ARPES experiment would be helpful in order to distinguish between the surface state and the bulk valence band. In the $k_{||}$-range of 0 to $0.12\,Å^{-1}$, the topological surface state lies below the Fermi level and should be detectable in ARPES. Further, STS at low temperature and in a magnetic field would be an appropriate technique in order to resolve the TI nature of the surface states by Landau level spectroscopy.

In the second part of the thesis, STM and STS measurements on the cleavage plane of single crystal $Bi_{14}Rh_3I_9$ have been carried out. This material is the first synthesized WTI and is stacked by consecutive graphene-like 2D TIs and insulating spacer layers. The cleavage plane is the topologically dark surface with the surrounding surfaces exhibiting topologically protected surface states. The different layers have been identified by atomically resolved STM. Further, a bias-dependent surface modulation of the 2D TI layer was found in excellent agreement with DFT, induced by the different stacking between the 2D TI layer and the underlying zigzag chain structure of the spacer layer.

The STS showed that 0.8 nm wide electron channels are present at surface step edges of the topologically dark surface. These electron channels have been found to be continuous in both energy and space within a large band gap of 200 meV, thereby, evidencing its non-trivial topology. The absence of these channels in the closely related, but topologically trivial insulator $Bi_{13}Pt_3I_7$ corroborates the channels' topological nature. The back-scatter-free electron channels are hereby a direct consequence of $Bi_{14}Rh_3I_9$'s structure, a stack of 2D TI, graphene-like planes separated by trivial insulators. It was further demonstrated that the surface of $Bi_{14}Rh_3I_9$ can be engraved using an AFM, allowing networks of protected channels to be patterned with nm precision. This might offer the opportunity to design spin filters [199] with extremely small footprint compared to 2D TIs in heterostructures [7]. Moreover, the interfacing with other materials such as superconductors or magnetic insulators required for advanced quantum circuitry [200, 201] becomes directly accessible by shadow mask evaporation. In this sense, the discovery of the first weak 3D TI $Bi_{14}Rh_3I_9$ might offer similar advantages as graphene does with respect to conventional semiconductor heterostructures [47].

Bibliography

[1] K. v. Klitzing, G. Dorda, and M. Pepper. New method for high-accuracy determination of the fine-structure constant based on quantized Hall resistance. *Phys. Rev. Lett.*, 45:494–497, Aug 1980.

[2] D. C. Tsui, H. L. Stormer, and A. C. Gossard. Two-dimensional magnetotransport in the extreme quantum limit. *Phys. Rev. Lett.*, 48:1559–1562, May 1982.

[3] D. J. Thouless, M. Kohmoto, M. P. Nightingale, and M. den Nijs. Quantized Hall conductance in a two-dimensional periodic potential. *Phys. Rev. Lett.*, 49:405–408, Aug 1982.

[4] Xiao-Gang Wen. Topological orders and edge excitations in fractional quantum Hall states. *Advances in Physics*, 44(5):405–473, 1995.

[5] B. Andrei Bernevig, Taylor L. Hughes, and Shou-Cheng Zhang. Quantum spin Hall effect and topological phase transition in HgTe quantum wells. *Science*, 314(5806):1757–1761, 2006.

[6] B. Andrei Bernevig and Shou-Cheng Zhang. Quantum spin Hall effect. *Phys. Rev. Lett.*, 96:106802, Mar 2006.

[7] Markus König, Steffen Wiedmann, Christoph Brüne, Andreas Roth, Hartmut Buhmann, Laurens W. Molenkamp, Xiao-Liang Qi, and Shou-Cheng Zhang. Quantum spin Hall insulator state in HgTe quantum wells. *Science*, 318(5851):766–770, 2007.

[8] C. L. Kane and E. J. Mele. Z_2 topological order and the quantum spin Hall effect. *Phys. Rev. Lett.*, 95:146802, Sep 2005.

[9] M. Z. Hasan and C. L. Kane. Colloquium: Topological insulators. *Rev. Mod. Phys.*, 82:3045–3067, 2010.

[10] X. L. Qi and S. C. Zhang. Topological insulators and superconductors. *Rev. Mod. Phys.*, 83:1057–1110, 2011.

[11] B. Yan and S. C. Zhang. Topological materials. *Rep. Prog. Phys.*, 75: 096501, 2012.

[12] Liang Fu and C. L. Kane. Time reversal polarization and a Z_2 adiabatic spin pump. *Phys. Rev. B*, 74:195312, Nov 2006.

[13] Liang Fu, C. L. Kane, and E. J. Mele. Topological insulators in three dimensions. *Phys. Rev. Lett.*, 98:106803, Mar 2007.

[14] Liang Fu and C. L. Kane. Topological insulators with inversion symmetry. *Phys. Rev. B*, 76:045302, Jul 2007.

[15] Zohar Ringel, Yaacov E. Kraus, and Ady Stern. Strong side of weak topological insulators. *Phys. Rev. B*, 86:045102, Jul 2012.

[16] Liang Fu and C. L. Kane. Topology, delocalization via average symmetry and the symplectic Anderson transition. *Phys. Rev. Lett.*, 109:246605, Dec 2012.

[17] Roger S. K. Mong, Jens H. Bardarson, and Joel E. Moore. Quantum transport and two-parameter scaling at the surface of a weak topological insulator. *Phys. Rev. Lett.*, 108:076804, Feb 2012.

[18] Koji Kobayashi, Tomi Ohtsuki, and Ken-Ichiro Imura. Disordered weak and strong topological insulators. *Phys. Rev. Lett.*, 110:236803, Jun 2013.

[19] Hideaki Obuse, Shinsei Ryu, Akira Furusaki, and Christopher Mudry. Spin-directed network model for the surface states of weak three-dimensional $\mathbb{Z}_2$ topological insulators. *Phys. Rev. B*, 89:155315, Apr 2014.

[20] Yukinori Yoshimura, Akihiko Matsumoto, Yositake Takane, and Ken-Ichiro Imura. Perfectly conducting channel on the dark surface of weak topological insulators. *Phys. Rev. B*, 88:045408, Jul 2013.

[21] Chao-Xing Liu, Xiao-Liang Qi, and Shou-Cheng Zhang. Half quantum spin Hall effect on the surface of weak topological insulators. *Physica E: Low-dimensional Systems and Nanostructures*, 44(5):906 – 911, 2012. ISSN 1386-9477. SI:Topological Insulators.

[22] S. LaShell, B. A. McDougall, and E. Jensen. Spin splitting of an Au(111) surface state band observed with angle resolved photoelectron spectroscopy. *Phys. Rev. Lett.*, 77:3419–3422, Oct 1996.

[23] M. Hoesch, M. Muntwiler, V. N. Petrov, M. Hengsberger, L. Patthey, M. Shi, M. Falub, T. Greber, and J. Osterwalder. Spin structure of the Shockley surface state on Au(111). *Phys. Rev. B*, 69:241401, Jun 2004.

[24] G. Jezequel, Y. Petroff, R. Pinchaux, and Félix Yndurain. Electronic structure of the Bi(111) surface. *Phys. Rev. B*, 33:4352–4355, Mar 1986.

[25] Christian R. Ast and Hartmut Höchst. Fermi surface of Bi(111) measured by photoemission spectroscopy. *Phys. Rev. Lett.*, 87:177602, Oct 2001.

[26] Yu. M. Koroteev, G. Bihlmayer, J. E. Gayone, E. V. Chulkov, S. Blügel, P. M. Echenique, and Ph. Hofmann. Strong spin-orbit splitting on Bi surfaces. *Phys. Rev. Lett.*, 93:046403, Jul 2004.

[27] A. Kimura, E. E. Krasovskii, R. Nishimura, K. Miyamoto, T. Kadono, K. Kanomaru, E. V. Chulkov, G. Bihlmayer, K. Shimada, H. Namatame, and M. Taniguchi. Strong Rashba-type spin polarization of the photocurrent from bulk continuum states: Experiment and theory for Bi(111). *Phys. Rev. Lett.*, 105:076804, Aug 2010.

[28] Christian R. Ast, Jürgen Henk, Arthur Ernst, Luca Moreschini, Mihaela C. Falub, Daniela Pacilé, Patrick Bruno, Klaus Kern, and Marco Grioni. Giant spin splitting through surface alloying. *Phys. Rev. Lett.*, 98:186807, May 2007.

[29] C. Pauly, G. Bihlmayer, M. Liebmann, M. Grob, A. Georgi, D. Subramaniam, M. R. Scholz, J. Sánchez-Barriga, A. Varykhalov, S. Blügel, O. Rader, and M. Morgenstern. Probing two topological surface bands of Sb_2Te_3 by spin-polarized photoemission spectroscopy. *Phys. Rev. B*, 86:235106, Dec 2012.

[30] Christian Pauly, Marcus Liebmann, Alessandro Giussani, Jens Kellner, Sven Just, Jaime Sanchez-Barriga, Emile Rienks, Oliver Rader, Raffaella Calarco, Gustav Bihlmayer, and Markus Morgenstern. Evidence for topological band inversion of the phase change material Ge2Sb2Te5. *Applied Physics Letters*, 103(24):243109, 2013.

[31] C. Pauly, C. Saunus, M. Liebmann, and M. Morgenstern. Spatially resolved Landau level spectroscopy of the topological Dirac cone of bulk-type Sb2Te3(0001): Potential fluctuations and quasiparticle lifetime. *Phys. Rev. B*, 92:085140, Aug 2015.

[32] B. Rasche, A. Isaeva, M. Ruck, S. Borisenko, V. Zabolotnyy, B. Büchner, K. Koepernik, C. Ortix, M. Richter, and J. van den Brink. Stacked topological insulator built from bismuth-based graphene sheet analogues. *Nature Mat.*, 12:422–425, 2013.

[33] Christian Pauly, Bertold Rasche, Klaus Koepernik, Marcus Liebmann, Marco Pratzer, Manuel Richter, Jens Kellner, Markus Eschbach, Bernhard Kaufmann, Lukasz Plucinski, Claus M. Schneider, Michael Ruck, Jeroen van den Brink, and Markus Morgenstern. Subnanometre-wide electron channels protected by topology. *Nature Phys.*, 11:338 – 343, 2015.

[34] Charles L. Kane and Eugene J. Mele. A new spin on the insulating state. *Science*, 314(5806):1692–1693, 2006.

[35] Walter Kohn. Theory of the insulating state. *Phys. Rev.*, 133:A171–A181, Jan 1964.

[36] R. B. Laughlin. Quantized Hall conductivity in two dimensions. *Phys. Rev. B*, 23:5632–5633, May 1981.

[37] M. Kohmoto. Topological invariant and the quantization of the Hall conductance. *Ann. Phys.*, 160:343–354, 1985.

[38] Watson G. Hall conductance as a topological invariant. *Contemporary Physics*, 37:127–143, 1996.

[39] G. D. Mahan. *Many-particle Physics*. Kluwer Academic Publishers, 3 edition, 2000.

[40] J. E. Avron, R. Seiler, and B. Simon. Homotopy and quantization in condensed matter physics. *Phys. Rev. Lett.*, 51:51–53, Jul 1983.

[41] Yasuhiro Hatsugai. Chern number and edge states in the integer quantum Hall effect. *Phys. Rev. Lett.*, 71:3697–3700, Nov 1993.

[42] C. L. Kane and E. J. Mele. Quantum spin Hall effect in graphene. *Phys. Rev. Lett.*, 95:226801, Nov 2005.

[43] P. W. Anderson. Absence of diffusion in certain random lattices. *Phys. Rev.*, 109:1492–1505, Mar 1958.

[44] D. Belitz and T. R. Kirkpatrick. The Anderson-Mott transition. *Rev. Mod. Phys.*, 66:261–380, Apr 1994.

[45] Kentaro Nomura, Mikito Koshino, and Shinsei Ryu. Topological delocalization of two-dimensional massless Dirac fermions. *Phys. Rev. Lett.*, 99:146806, Oct 2007.

[46] K. S. Novoselov, A. K. Geim, S. V. Morozov, D. Jiang, M. I. Katsnelson, I. V. Grigorieva, S. V. Dubonos, and A. A. Firsov. Two-dimensional gas of massless Dirac fermions in graphene. *Nature*, 438:197–200, 2005.

[47] A. H. Castro Neto, F. Guinea, N. M. R. Peres, K. S. Novoselov, and A. K. Geim. The electronic properties of graphene. *Rev. Mod. Phys.*, 81:109–162, Jan 2009.

[48] M. Gmitra, S. Konschuh, C. Ertler, C. Ambrosch-Draxl, and J. Fabian. Band-structure topologies of graphene: Spin-orbit coupling effects from first principles. *Phys. Rev. B*, 80:235431, Dec 2009.

[49] J. E. Moore and L. Balents. Topological invariants of time-reversal-invariant band structures. *Phys. Rev. B*, 75:121306, Mar 2007.

[50] Takahiro Fukui and Yasuhiro Hatsugai. Quantum spin Hall effect in three dimensional materials: Lattice computation of Z2 topological invariants and its application to Bi and Sb. *Journal of the Physical Society of Japan*, 76(5):053702, 2007.

[51] Xiao-Liang Qi, Taylor L. Hughes, and Shou-Cheng Zhang. Topological field theory of time-reversal invariant insulators. *Phys. Rev. B*, 78: 195424, Nov 2008.

[52] D. N. Sheng, Z. Y. Weng, L. Sheng, and F. D. M. Haldane. Quantum spin-Hall effect and topologically invariant Chern numbers. *Phys. Rev. Lett.*, 97:036808, Jul 2006.

[53] Rahul Roy. Z_2 classification of quantum spin Hall systems: An approach using time-reversal invariance. *Phys. Rev. B*, 79:195321, May 2009.

[54] Rahul Roy. Topological phases and the quantum spin Hall effect in three dimensions. *Phys. Rev. B*, 79:195322, May 2009.

[55] Zhong Wang, Xiao-Liang Qi, and Shou-Cheng Zhang. Equivalent topological invariants of topological insulators. *New Journal of Physics*, 12(6): 065007, 2010.

[56] S-Q. Shen. *Topological insulators: Dirac equation in condensed matters.* Springer Verlag, 2012.

[57] M. Morgenstern. Festkörperphysik ii: Wechselwirkungen. In *Vorlesungsskript.* RWTH Aachen, 2014.

[58] R. D. King-Smith and David Vanderbilt. Theory of polarization of crystalline solids. *Phys. Rev. B*, 47:1651–1654, Jan 1993.

[59] A. Cayley. On the theory of permutants. *Cambridge and Dublin Mathematical Journal VII*, 2:40–51, 1852.

[60] B. A. Bernevig and T. L. Hughes. *Topological insulators and topological superconductors.* Princeton University Press, 2013.

[61] Jing-Min Hou, Wen-Xin Zhang, and Guo-Xiang Wang. Three-dimensional topological insulators in the octahedron-decorated cubic lattice. *Phys. Rev. B*, 84:075105, Aug 2011.

[62] Binghai Yan, Lukas Müchler, and Claudia Felser. Prediction of weak topological insulators in layered semiconductors. *Phys. Rev. Lett.*, 109: 116406, Sep 2012.

[63] Peizhe Tang, Binghai Yan, Wendong Cao, Shu-Chun Wu, Claudia Felser, and Wenhui Duan. Weak topological insulators induced by the interlayer coupling: A first-principles study of stacked Bi_2TeI. *Phys. Rev. B*, 89:041409, Jan 2014.

[64] Gang Yang, Junwei Liu, Liang Fu, Wenhui Duan, and Chaoxing Liu. Weak topological insulators in PbTe/SnTe superlattices. *Phys. Rev. B*, 89:085312, Feb 2014.

[65] D. Hsieh, D. Qian, L. Wray, Y. Xia, Y. S. Hor, R. J. Cava, and M. Z. Hasan. A topological Dirac insulator in a quantum spin Hall phase. *Nature*, 452: 970–974, 2008.

[66] D. Hsieh, Y. Xia, L. Wray, D. Qian, A. Pal, J. H. Dil, J. Osterwalder, F. Meier, G. Bihlmayer, C. L. Kane, Y. S. Hor, R. J. Cava, and M. Z. Hasan. Observation of unconventional quantum spin textures in topological insulators. *Science*, 323(5916):919–922, 2009.

[67] Haijun Zhang, Chao-Xing Liu, Xiao-Liang Qi, Xi Dai, Zhong Fang, and Shou-Cheng Zhang. Topological insulators in Bi2Se3, Bi2Te3 and Sb2Te3 with a single Dirac cone on the surface. *Nature Phys.*, 5:438–442, 2009.

[68] Y. Xia, D. Qian, D. Hsieh, L. Wray, A. Pal, H. Lin, A. Bansil, D. Grauer, Y. S. Hor, R. J. Cava, and M. Z. Hasan. Observation of a large-gap topological-insulator class with a single Dirac cone on the surface. *Nature Phys.*, 5:398–402, 2009.

[69] D. Hsieh, Y. Xia, D. Qian, L. Wray, J. H. Dil, F. Meier, J. Osterwalder, L. Patthey, J. G. Checkelsky, N. P. Ong, A. V. Fedorov, H. Lin, A. Bansil, D. Grauer, Y. S. Hor, R. J. Cava, and M. Z. Hasan. A tunable topological insulator in the spin helical Dirac transport regime. *Nature*, 460:1101–1105, 2009.

[70] Y. L. Chen, J.-H. Chu, J. G. Analytis, Z. K. Liu, K. Igarashi, H.-H. Kuo, X. L. Qi, S. K. Mo, R. G. Moore, D. H. Lu, M. Hashimoto, T. Sasagawa, S. C. Zhang, I. R. Fisher, Z. Hussain, and Z. X. Shen. Massive Dirac fermion on the surface of a magnetically doped topological insulator. *Science*, 329(5992):659–662, 2010.

[71] Y. L. Chen, J. G. Analytis, J.-H. Chu, Z. K. Liu, S.-K. Mo, X. L. Qi, H. J. Zhang, D. H. Lu, X. Dai, Z. Fang, S. C. Zhang, I. R. Fisher, Z. Hussain, and Z.-X. Shen. Experimental realization of a three-dimensional topological insulator, Bi2Te3. *Science*, 325(5937):178–181, 2009.

[72] D. Hsieh, Y. Xia, D. Qian, L. Wray, F. Meier, J. H. Dil, J. Osterwalder, L. Patthey, A. V. Fedorov, H. Lin, A. Bansil, D. Grauer, Y. S. Hor, R. J. Cava, and M. Z. Hasan. Observation of time-reversal-protected single-Dirac-cone topological-insulator states in Bi_2Te_3 and Sb_2Te_3. *Phys. Rev. Lett.*, 103:146401, Sep 2009.

[73] Yeping Jiang, Y. Y. Sun, Mu Chen, Yilin Wang, Zhi Li, Canli Song, Ke He, Lili Wang, Xi Chen, Qi-Kun Xue, Xucun Ma, and S. B. Zhang. Fermi-level tuning of epitaxial Sb_2Te_3 thin films on graphene by regulating intrinsic defects and substrate transfer doping. *Phys. Rev. Lett.*, 108: 066809, Feb 2012.

[74] Sunghun Kim, M. Ye, K. Kuroda, Y. Yamada, E. E. Krasovskii, E. V. Chulkov, K. Miyamoto, M. Nakatake, T. Okuda, Y. Ueda, K. Shimada, H. Namatame, M. Taniguchi, and A. Kimura. Surface scattering via bulk continuum states in the 3D topological insulator Bi_2Se_3. *Phys. Rev. Lett.*, 107:056803, Jul 2011.

[75] Z.-H. Pan, E. Vescovo, A. V. Fedorov, D. Gardner, Y. S. Lee, S. Chu, G. D. Gu, and T. Valla. Electronic structure of the topological insulator Bi_2Se_3 using Angle-Resolved Photoemission Spectroscopy: Evidence for a nearly full surface spin polarization. *Phys. Rev. Lett.*, 106: 257004, Jun 2011.

[76] L. Wray, S.-Y. Xu, Y. Xia, D. Hsieh, A. V. Fedorov, Y. S. Hor, R. J. Cava, A. Bansil, H. Lin, and M. Z. Hasan. A topological insulator surface under strong Coulomb, magnetic and disorder perturbations. *Nature Phys.*, 7:32–37, 2010.

[77] S.-Y. Xu, M. Neupane, C. Liu, D. Zhang, A. Richardella, L. Wray, N. Alidoust, M. Leandersson, T. Balasubramaniam, J. Sanchez-Barriga, O. Rader, G. Landolt, B. Slomski, J. H. Dil, J. Osterwalder, T-R. Chang, H-T. Jeng, H. Lin, A. Bansil, N. Samarth, and M. Z. Hasan. Hedgehog

spin texture and Berrys phase tuning in a magnetic topological insulator. *Nature Phys.*, 8:616–622, 2012.

[78] Tong Zhang, Peng Cheng, Xi Chen, Jin-Feng Jia, Xucun Ma, Ke He, Lili Wang, Haijun Zhang, Xi Dai, Zhong Fang, Xincheng Xie, and Qi-Kun Xue. Experimental demonstration of topological surface states protected by time-reversal symmetry. *Phys. Rev. Lett.*, 103:266803, Dec 2009.

[79] Peng Cheng, Canli Song, Tong Zhang, Yanyi Zhang, Yilin Wang, Jin-Feng Jia, Jing Wang, Yayu Wang, Bang-Fen Zhu, Xi Chen, Xucun Ma, Ke He, Lili Wang, Xi Dai, Zhong Fang, Xincheng Xie, Xiao-Liang Qi, Chao-Xing Liu, Shou-Cheng Zhang, and Qi-Kun Xue. Landau quantization of topological surface states in Bi_2Se_3. *Phys. Rev. Lett.*, 105:076801, Aug 2010.

[80] P. Roushan, J. Seo, C. V. Parker, Y. S. Hor, D. Hsieh, D. Qian, A. Richardella, M. Z. Hasan, R. J. Cava, and A. Yazdani. Topological surface states protected from backscattering by chiral spin texture. *Nature*, 460:1106–1109, 2009.

[81] Zhanybek Alpichshev, J. G. Analytis, J.-H. Chu, I. R. Fisher, Y. L. Chen, Z. X. Shen, A. Fang, and A. Kapitulnik. STM imaging of electronic waves on the surface of Bi_2Te_3: topologically protected surface states and hexagonal warping effects. *Phys. Rev. Lett.*, 104:016401, Jan 2010.

[82] H. Beidenkopf, P. Roushan, J. Seo, L. Gorman, I. Drozdov, Y. S. Hor, R. J. Cava, and A. Yazdani. Spatial fluctuations of helical dirac fermions on the surface of topological insulators. *Nature Phys.*, 7:939–943, 2011.

[83] T. Hanaguri, K. Igarashi, M. Kawamura, H. Takagi, and T. Sasagawa. Momentum-resolved Landau-level spectroscopy of Dirac surface state in Bi_2Se_3. *Phys. Rev. B*, 82:081305, Aug 2010.

[84] Yeping Jiang, Yilin Wang, Mu Chen, Zhi Li, Canli Song, Ke He, Lili Wang, Xi Chen, Xucun Ma, and Qi-Kun Xue. Landau quantization and the thickness limit of topological insulator thin films of Sb_2Te_3. *Phys. Rev. Lett.*, 108:016401, Jan 2012.

[85] J. Seo, P. Roushan, H. Beidenkopf, Y. S. Hor, R. J. Cava, and A. Yazdani. Transmission of topological surface states through surface barriers. *Nature*, 466:343–346, 2010.

[86] Liang Fu. Topological crystalline insulators. *Phys. Rev. Lett.*, 106:106802, Mar 2011.

[87] P. Dziawa, B. J. Kowalski, K. Dybko, R. Buczko, A. Szczerbakow, M. Szot, E. Lusakowska, T. Balasubramaniam, B. M. Wojek, M. H. Berntsen, O. Tjernberg, and T. Story. Topological crystalline insulator states in $Pb_{1-x}Sn_xSe$. *Nature Mat.*, 11:1023–1027, 2012.

[88] Rui Yu, Wei Zhang, Hai-Jun Zhang, Shou-Cheng Zhang, Xi Dai, and Zhong Fang. Quantized anomalous Hall effect in magnetic topological insulators. *Science*, 329(5987):61–64, 2010.

[89] Cui-Zu Chang, Jinsong Zhang, Xiao Feng, Jie Shen, Zuocheng Zhang, Minghua Guo, Kang Li, Yunbo Ou, Pang Wei, Li-Li Wang, Zhong-Qing Ji, Yang Feng, Shuaihua Ji, Xi Chen, Jinfeng Jia, Xi Dai, Zhong Fang, Shou-Cheng Zhang, Ke He, Yayu Wang, Li Lu, Xu-Cun Ma, and Qi-Kun Xue. Experimental observation of the quantum anomalous Hall effect in a magnetic topological insulator. *Science*, 340(6129):167–170, 2013.

[90] G. Binnig and H. Rohrer. Rastertunnelmikroskopie. *Helv. Phys. Acta*, 55:726, 1982.

[91] R. J. Hamers. Atomic-resolution surface spectroscopy with the scanning tunneling microscope. *Annu. Rev. Phys. Chem.*, 40:531, 1989.

[92] J. Tersoff and D. R. Hamann. Theory of the scanning tunneling microscope. *Phys. Rev. B*, 31:805–813, 1985.

[93] J. Bardeen. Tunnelling from a many-particle point of view. *Phys. Rev. Lett.*, 6:57–59, 1961.

[94] J. Tersoff and D. R. Hamann. Theory and application for the scanning tunneling microscope. *Phys. Rev. Lett.*, 50:1998–2001, 1983.

[95] C. Julian Chen. Origin of atomic resolution on metal surfaces in scanning tunneling microscopy. *Phys. Rev. Lett.*, 65:448–451, Jul 1990.

[96] J. Wintterlin, J. Wiechers, H. Brune, T. Gritsch, H. Höfer, and R. J. Behm. Atomic-resolution imaging of close-packed metal surfaces by scanning tunneling microscopy. *Phys. Rev. Lett.*, 62:59–62, Jan 1989.

[97] V. M. Hallmark, S. Chiang, J. F. Rabolt, J. D. Swalen, and R. J. Wilson. Observation of atomic corrugation on Au(111) by scanning tunneling microscopy. *Phys. Rev. Lett.*, 59:2879–2882, Dec 1987.

[98] A. Selloni, P. Carnevali, E. Tosatti, and C. D. Chen. Voltage-dependent STM of a crystal surface. *Graph. Rev. B.*, 31:2602, 1985.

[99] A. Wachowiak. *Aufbau einer 300mK-Ultrahochvakuum-Rastertunnelmikroskopie-Anlage mit 14 Tesla Magnet und spinpolarisierte Rastertunnelspektroskopie an ferromagnetischen Fe-Inseln.* PhD thesis, Universität Hamburg, 2003.

[100] M. Morgenstern. Probing the local density of states of dilute electron systems in different dimensions. *Surf. Rev. Lett.*, 10:933–962, 2003.

[101] Andrea Damascelli. Probing the electronic structure of complex systems by ARPES. *Physica Scripta*, 2004(T109):61, 2004.

[102] C. N. Berglund and W. E. Spicer. Photoemission studies of copper and silver: Theory. *Phys. Rev.*, 136:A1030–A1044, Nov 1964.

[103] S. Hüfner. *Photoelectron spectroscopy.* Springer-Verlag, Berlin, 1995.

[104] M. P. Seah and W. A. Dench. Quantitative electron spectroscopy of surfaces: A standard data base for electron inelastic mean free paths in solids. *Surface and Interface Analysis*, 1(1):2–11, 1979. ISSN 1096-9918.

[105] Peter J. Feibelman and D. E. Eastman. Photoemission spectroscopy-correspondence between quantum theory and experimental phenomenology. *Phys. Rev. B*, 10:4932–4947, Dec 1974.

[106] H. Y. Fan. Theory of photoelectric emission from metals. *Phys. Rev.*, 68: 43–52, Jul 1945.

[107] Andrea Damascelli, Zahid Hussain, and Zhi-Xun Shen. Angle-resolved photoemission studies of the cuprate superconductors. *Rev. Mod. Phys.*, 75:473–541, Apr 2003.

[108] E.-E. Koch, G. V. Marr, G.S. Brown, D. E. Moncton, S. Ebashi, M. Koch, and E. Rubenstein. *Handbook on Synchrotron Radiation.* Elsevier Science, Amsterdam, 191.

[109] G. Beamson, D. Briggs, S. F. Davies, I. W. Fletcher, D. T. Clark, J. Howard, U. Gelius, B. Wannberg, and P. Balzer. Performance and application of the scienta ESCA300 spectrometer. *Surface and Interface Analysis*, 15(9):541–549, 1990. ISSN 1096-9918.

[110] N. Martensson, P. Baltzer, P.A. Brühwiler, J.-O. Forsell, A. Nilsson, A. Stenborg, and B. Wannberg. A very high resolution electron spectrometer. *Journal of Electron Spectroscopy and Related Phenomena*, 70(2): 117 – 128, 1994. ISSN 0368-2048.

[111] M. Wuttig and N. Yamada. Phase-change materials for rewritable data storage. *Nat. Mater.*, 6:824–832, 2007.

[112] Matthias Wuttig and Simone Raoux. The science and technology of phase change materials. *Zeitschrift für anorganische und allgemeine Chemie*, 638(15):2455–2465, 2012. ISSN 1521-3749.

[113] Stanford R. Ovshinsky. Reversible electrical switching phenomena in disordered structures. *Phys. Rev. Lett.*, 21:1450–1453, Nov 1968.

[114] Noboru Yamada, Eiji Ohno, Kenichi Nishiuchi, Nobuo Akahira, and Masatoshi Takao. Rapid-phase transitions of GeTe-Sb2Te3 pseudobinary amorphous thin films for an optical disk memory. *Journal of Applied Physics*, 69(5):2849–2856, 1991.

[115] M. H. R. Lankhorst, B. W. S. M. M. Ketelaars, and R. A. M. Wolters. Low-cost and nanoscale non-volatile memory concept for future silicon chips. *Nat. Mater.*, 4:347–352, 2005.

[116] M. Wuttig. Phase-change materials: Towards a universal memory? *Nature Mat.*, 4:265–266, 2005.

[117] D. Lencer, M. Salinga, B. Grabowski, T. Hickel, J. Neugebauer, and M. Wuttig. A map for phase-change materials. *Nature Mat.*, 7:972–977, 2008.

[118] K. Shportko, S. Kremers, M. Woda, D. Lencer, J. Robertson, and M. Wuttig. Resonant bonding in crystalline phase-change materials. *Nat. Mater.*, 7:653–658, 2008.

[119] S. Eremeev, G. Landolt, T. V. Menshchikova, B. Slomski, Y. Koroteev, Z. Aliev, M. Babanly, J. Henk, A. Ernst, L. Patthey, A. Eich, A. Khatjetoorians, J. Hagemeister, O. Pietzsch, J. Wiebe, R. Wiesendanger, P. M. Echenique, S. Tsirkin, I. Amiraslanov, J. H. Dil, and E. Chulkov. Atom-specific spin mapping and buried topological states in a homologous series of topological insulators. *Nat. Commun.*, 3:635, 2012.

[120] Jeongwoo Kim, Jinwoong Kim, and Seung-Hoon Jhi. Prediction of topological insulating behavior in crystalline Ge-Sb-Te. *Phys. Rev. B*, 82:201312, Nov 2010.

[121] Jeongwoo Kim, Jinwoong Kim, Ki-Seok Kim, and Seung-Hoon Jhi. Topological phase transition in the interaction of surface Dirac fermions in heterostructures. *Phys. Rev. Lett.*, 109:146601, Oct 2012.

[122] Jeongwoo Kim and Seung-Hoon Jhi. Emerging topological insulating phase in Ge-Sb-Te compounds. *physica status solidi (b)*, 249(10):1874–1879, 2012. ISSN 1521-3951.

[123] I.V. Silkin, Yu.M. Koroteev, G. Bihlmayer, and E.V. Chulkov. Influence of the Ge-Sb sublattice atomic composition on the topological electronic properties of Ge2Sb2Te5. *Applied Surface Science*, 267(0):169 – 172, 2013. ISSN 0169-4332. 11th International Conference on Atomically Controlled Surfaces, Interfaces and Nanostructures.

[124] D. Loke, T. H. Lee, W. J. Wang, L. P. Shi, R. Zhao, Y. C. Yeo, T. C. Chong, and S. R. Elliott. Breaking the speed limits of phase-change memory. *Science*, 336(6088):1566–1569, 2012.

[125] Feng Xiong, Albert D. Liao, David Estrada, and Eric Pop. Low-power switching of phase-change materials with carbon nanotube electrodes. *Science*, 332(6029):568–570, 2011.

[126] W. Welnic. *Electronic and optical properties of phase change alloys studied with ab initio methods*. PhD thesis, RWTH Aachen, 2002.

[127] M. Chen, K. A. Rubin, and R. W. Barton. Compound materials for reversible, phase-change optical data storage. *Applied Physics Letters*, 49 (9):502–504, 1986.

[128] N. Yamada, M. Takao, and M. Takenaga. Te-Ge-Sn-Au phase change recording film for optical disk. *Proc. SPIE*, 0695:79–85, 1987.

[129] Eiji Ohno, Noboru Yamada, Toshimitsu Kurumizawa, Kunio Kimura, and Masatoshi Takao. TeGeSnAu alloys for phase change type optical disk memories. *Japanese Journal of Applied Physics*, 28(7R):1235, 1989.

[130] Wojciech Welnic, Silvana Botti, Lucia Reining, and Matthias Wuttig. Origin of the optical contrast in phase-change materials. *Phys. Rev. Lett.*, 98:236403, Jun 2007.

[131] J-B. Park, G-S. Park, H-S. Baik, J-H. Lee, H. Jeong, and K. Kim. Phase-change behavior of stoichiometric Ge2Sb2Te5 in phase-change random access memory. *J. Electrochem. Soc.*, 154:H139, 2007.

[132] T. Matsunaga and N. Yamada. Structural investigation of GeSb2Te4: A high-speed phase-change material. *Phys. Rev. B*, 69:104111, 2004.

[133] M Wuttig, D. Lüsebrink, D. Wamwangi, W. Welnic, M. Gillessen, and R. Dronskowski. The role of vacancies and local distortions in the design of new phase-change materials. *Nature Mat.*, 6:122–128, 2007.

[134] J. Akola and R. O. Jones. Structural phase transitions on the nanoscale: The crucial pattern in the phase-change materials $Ge_2Sb_2Te_5$ and GeTe. *Phys. Rev. B*, 76:235201, Dec 2007.

[135] T. Siegrist, P. Jost, M. Volker, M. Woda, P. Merkelbach, C. Schlock-ermann, and M. Wuttig. Disorder-induced localization in crystalline phase-change materials. *Nature Mat.*, 10:202–208, 2011.

[136] A. V. Kolobov, P. Fons, A. I. Frenkel, A. L. Ankudinov, J. Tominaga, and T. Uruga. Understanding the phase-change mechanism of rewritable optical media. *Nature Mat.*, 3:703–708, 2004.

[137] W. Welnic, A. Pamungkas, R. Detemple, C. Steimer, Blügel S., and M. Wuttig. Unravelling the interplay of local structure and physical properties in phase-change materials. *Nature Mat..*, 5:56–62, 2006.

[138] Toshiyuki Matsunaga, Noboru Yamada, and Yoshiki Kubota. Structures of stable and metastable Ge2Sb2Te5, an intermetallic compound in GeTe-Sb2Te3 pseudobinary systems. *Acta Crystallographica Section B*, 60(6):685–691, Dec 2004.

[139] I. I. Petrov, R. M. Imamov, and Z. G. Pinsker. Electron-diffraction determination of structures of Ge2Sb2Te5 and GeSb4Te7. *Sov. Phys. Crystallogr.*, 13:339, 1968.

[140] B. J. Kooi and J. Th. M. De Hosson. Electron diffraction and high-resolution transmission electron microscopy of the high temperature crystal structures of GexSb2Te3+x(x=1,2,3) phase change material. *Journal of Applied Physics*, 92(7):3584–3590, 2002.

[141] Jino Im, Jae-Hyeon Eom, Changwon Park, Kimin Park, Dong-Seok Suh, Kijoon Kim, Youn-Seon Kang, Cheolkyu Kim, Tae-Yon Lee, Yoonho Khang, Young-Gui Yoon, and Jisoon Ihm. Hierarchical structure and phase transition of $(GeTe)_n (Sb_2Te_3)_m$ used for phase-change memory. *Phys. Rev. B*, 78:205205, Nov 2008.

[142] Walter K. Njoroge, Han-Willem Wöltgens, and Matthias Wuttig. Density changes upon crystallization of Ge2Sb2.04Te4.74 films. *J. Vac. Sci. Technol. A*, 20(1):230–233, 2002.

[143] J. Kalb, F. Spaepen, and M. Wuttig. Atomic force microscopy measurements of crystal nucleation and growth rates in thin films of amorphous Te alloys. *Applied Physics Letters*, 84(25):5240–5242, 2004.

[144] D. Subramaniam, C. Pauly, M. Liebmann, M. Woda, P. Rausch, P. Merkelbach, M. Wuttig, and M. Morgenstern. Scanning tunneling microscopy and spectroscopy of the phase change alloy $Ge_1Sb_2Te_4$. *Applied Physics Letters*, 95(10):103110, 2009.

[145] I. Friedrich, V. Weidenhof, W. Njoroge, P. Franz, and M. Wuttig. Structural transformations of Ge2Sb2Te5 films studied by electrical resistance measurements. *Journal of Applied Physics*, 87(9):4130–4134, 2000.

[146] Jung-Jin Kim, Keisuke Kobayashi, Eiji Ikenaga, Masaaki Kobata, Shigenori Ueda, Toshiyuki Matsunaga, Kouichi Kifune, Rie Kojima, and Noboru Yamada. Electronic structure of amorphous and crystalline $(GeTe)_{1-x}(Sb_2Te_3)_x$ investigated using hard x-ray photoemission spectroscopy. *Phys. Rev. B*, 76:115124, Sep 2007.

[147] Guang Wang, Xiegang Zhu, Jing Wen, Xi Chen, Ke He, Lili Wang, Xucun Ma, Ying Liu, Xi Dai, Zhong Fang, Jinfeng Jia, and Qikun Xue. Atomically smooth ultrathin films of topological insulator Sb2Te3. *Nano Research*, 3(12):874–880, 2010. ISSN 1998-0124.

[148] T. Mashoff. *Design of a low-temperature scanning tunnelling microscope system used to examine graphene nanomembranes*. PhD thesis, RWTH Aachen University, 2010.

[149] T. Mashoff, M. Pratzer, and M. Morgenstern. A low-temperature high resolution scanning tunneling microscope with a three-dimensional magnetic vector field operating in ultrahigh vacuum. *Review of Scientific Instruments*, 80(5):053702, 2009.

[150] Collaboration: Authors and editors of the volumes III/17E-17F-41C. Antimony telluride (Sb2Te3) crystal structure, chemical bond, lattice parameters (including data for Bi2Se3, Bi2Te3). In O. Madelung, U. Rössler, and M. Schulz, editors, *Non-Tetrahedrally Bonded Elements and Binary Compounds I*, volume 41C of *Landolt-Börnstein - Group III Condensed Matter*, pages 1–4. Springer Berlin Heidelberg, 1998. ISBN 978-3-540-64583-2.

[151] John P. Perdew, Kieron Burke, and Matthias Ernzerhof. Generalized Gradient Approximation made simple. *Phys. Rev. Lett.*, 77:3865–3868, Oct 1996.

[152] Chun Li, A. J. Freeman, H. J. F. Jansen, and C. L. Fu. Magnetic anisotropy in low-dimensional ferromagnetic systems: Fe monolayers on Ag(001), Au(001), and Pd(001) substrates. *Phys. Rev. B*, 42:5433–5442, Sep 1990.

[153] S.V. Eremeev, Yu.M. Koroteev, and E.V. Chulkov. Effect of the atomic composition of the surface on the electron surface states in topological insulators $A_2^V B_3^{VI}$. *JETP Letters*, 91(8):387–391, 2010. ISSN 0021-3640.

[154] Guohong Li, Adina Luican, and Eva Y. Andrei. Scanning tunneling spectroscopy of graphene on graphite. *Phys. Rev. Lett.*, 102:176804, Apr 2009.

[155] M. Morgenstern, D. Haude, V. Gudmundsson, Chr. Wittneven, R. Dombrowski, Chr. Steinebach, and R. Wiesendanger. Low temperature scanning tunneling spectroscopy on InAs(110). *Journal of Electron Spectroscopy and Related Phenomena*, 109(1–2):127 – 145, 2000. ISSN 0368-2048.

[156] Su-Yang Xu, Y. Xia, L. A. Wray, S. Jia, F. Meier, J. H. Dil, J. Osterwalder, B. Slomski, A. Bansil, H. Lin, R. J. Cava, and M. Z. Hasan. Topological phase transition and texture inversion in a tunable topological insulator. *Science*, 332(6029):560–564, 2011.

[157] S. H. Pan, E. W. Hudson, and J. C. Davis. 3He refrigerator based very low temperature scanning tunneling microscope. *Rev. Sci. Instrum.*, 70: 1459, 1999.

[158] D. Kong, Y. Chen, J. J. Cha, Q. Zhang, J. G. Analytis, K. Lai, Z. Liu, S.S. Hong, K. J. Koski, S. K. Mo, Z. Hussain, I. R. Fisher, Z. X. Shen, and Y. Cui. Ambipolar field effect in the ternary topological insulator $(Bi_x Sb_{1-x})_2 Te_3$ by composition tuning. *Nature Nano.*, 6:705–709, 2011.

[159] Oleg V. Yazyev, Joel E. Moore, and Steven G. Louie. Spin polarization and transport of surface states in the topological insulators $Bi_2 Se_3$ and $Bi_2 Te_3$ from first principles. *Phys. Rev. Lett.*, 105:266806, Dec 2010.

[160] K. Ishizaka, M.S. Bahramy, H. Murakawa, M. Sakano, T. Shimojima, T. Sonobe, K. Koizumi, S. Shin, Miyahara H., A. Kimura, K. Miyamoto, T. Okuda, H. Namatame, M. Taniguchi, R. Arita, N. Nagaosa, K. Kobayashi, Y. Murakami, R. Kumai, Y. Kaneko, Y. Onose, and Y. Tokura. Giant Rashba-type spin splitting in bulk BiTeI. *Nature Mat.*, 10:521–526, 2011.

[161] J.B. Pendry and S.J. Gurman. Theory of surface states: General criteria for their existence. *Surface Science*, 49(1):87 – 105, 1975. ISSN 0039-6028.

[162] R. Feder and K. Sturm. Spin-orbit effects in the electronic structure of the (001) surface of bcc $4d$ and $5d$ transition metals. *Phys. Rev. B*, 12: 537–548, Jul 1975.

[163] R.F. Willis, B. Feuerbacher, and B. Fitton. Angular dependence of photoemission from intrinsic surface states on W(100). *Solid State Commu-
nications*, 18(9–10):1315 – 1319, 1976. ISSN 0038-1098.

[164] Shin Yaginuma, Katsumi Nagaoka, Tadaaki Nagao, Gustav Bihlmayer,
Yury M. Koroteev, Eugene V. Chulkov, and Tomonobu Nakayama.
Electronic structure of ultrathin bismuth films with A7 and black-
phosphorus-like structures. *Journal of the Physical Society of Japan*, 77
(1):014701, 2008.

[165] X. Gonze, J.-P. Michenaud, and J.-P. Vigneron. First-principles study of
As, Sb, and Bi electronic properties. *Phys. Rev. B*, 41:11827–11836, Jun
1990.

[166] S. Souma, M. Komatsu, M. Nomura, T. Sato, A. Takayama, T. Taka-
hashi, K. Eto, Kouji Segawa, and Yoichi Ando. Spin polarization of
gapped Dirac surface states near the topological phase transition in
$TlBi(S_{1-x}Se_x)_2$. *Phys. Rev. Lett.*, 109:186804, Nov 2012.

[167] Noboru Yamada and Toshiyuki Matsunaga. Structure of laser-
crystallized Ge2Sb2+xTe5 sputtered thin films for use in optical memory. *Journal of Applied Physics*, 88(12):7020–7028, 2000.

[168] Zhimei Sun, Jian Zhou, and Rajeev Ahuja. Structure of phase change
materials for data storage. *Phys. Rev. Lett.*, 96:055507, Feb 2006.

[169] Y. J. Park, J. Y. Lee, M. S. Youm, Y. T. Kim, and H. S. Lee. Crystal structure and atomic arrangement of the metastable Ge2Sb2Te5 thin films
deposited on SiO2/Si substrates by sputtering method. *Journal of Applied Physics*, 97(9):093506, 2005.

[170] Baisheng Sa, Jian Zhou, Zhitang Song, Zhimei Sun, and Rajeev Ahuja.
Pressure-induced topological insulating behavior in the ternary chalcogenide $Ge_2Sb_2Te_5$. *Phys. Rev. B*, 84:085130, Aug 2011.

[171] B. Sa, J. Zhou, Z. Sun, and R. Ahuja. Strain-induced topological insulating behavior in ternary chalcogenide Ge2Sb2Te5. *EPL (Europhysics
Letters)*, 97(2):27003, 2012.

[172] Ferhat Katmis, Raffaella Calarco, Karthick Perumal, Peter Rodenbach,
Alessandro Giussani, Michael Hanke, Andre Proessdorf, Achim Tram-
pert, Frank Grosse, Roman Shayduk, Richard Campion, Wolfgang
Braun, and Henning Riechert. Insight into the growth and control of
single-crystal layers of Ge-Sb-Te phase-change material. *Crystal Growth
& Design*, 11(10):4606–4610, 2011.

[173] Peter Rodenbach, Raffaella Calarco, Karthick Perumal, Ferhat Katmis, Michael Hanke, Andre Proessdorf, Wolfgang Braun, Alessandro Giussani, Achim Trampert, Henning Riechert, Paul Fons, and Alexander V. Kolobov. Epitaxial phase-change materials. *physica status solidi (RRL) - Rapid Research Letters*, 6(11):415–417, 2012. ISSN 1862-6270.

[174] Y. Takagaki, A. Giussani, K. Perumal, R. Calarco, and K.-J. Friedland. Robust topological surface states in Sb_2Te_3 layers as seen from the weak antilocalization effect. *Phys. Rev. B*, 86:125137, Sep 2012.

[175] Alessandro Giussani, Karthick Perumal, Michael Hanke, Peter Rodenbach, Henning Riechert, and Raffaella Calarco. On the epitaxy of germanium telluride thin films on silicon substrates. *physica status solidi (b)*, 249(10):1939–1944, 2012. ISSN 1521-3951.

[176] Y Takagaki, A Giussani, J Tominaga, U Jahn, and R Calarco. Transport properties in a Sb-Te binary topological-insulator system. *Journal of Physics: Condensed Matter*, 25(34):345801, 2013.

[177] R. Mazzarello and R Calarco. private communication. *not published*, 2014.

[178] Zheng Zhang, Jisheng Pan, Yong Lim Foo, Lina Wei-Wei Fang, Yee-Chia Yeo, Rong Zhao, Luping Shi, and Tow-Chong Chong. Effective method for preparation of oxide-free Ge2Sb2Te5 surface: An X-ray photoelectron spectroscopy study. *Applied Surface Science*, 256(24):7696 – 7699, 2010. ISSN 0169-4332.

[179] J. Kellner. Scanning tunneling microscopy investigation on phase change materials. Master's thesis, RWTH Aachen University, 2013.

[180] J. A. Bearden and A. F. Burr. Reevaluation of X-ray atomic energy levels. *Rev. Mod. Phys.*, 39:125–142, Jan 1967.

[181] John C. Fuggle and Nils Martensson. Core-level binding energies in metals. *Journal of Electron Spectroscopy and Related Phenomena*, 21(3):275 – 281, 1980. ISSN 0368-2048.

[182] Susan Mroczkowski and David Lichtman. Calculated Auger yields and sensitivity factors for KLL-NOO transitions with 1-10 kV primary beams. *J. Vac. Sci. Technol. A*, 3(4):1860–1865, 1985.

[183] B. J. Kooi, W. M. G. Groot, and J. Th. M. De Hosson. In situ transmission electron microscopy study of the crystallization of Ge2Sb2Te5. *Journal of Applied Physics*, 95(3):924–932, 2004.

[184] Toshihisa Nonaka, Gentaro Ohbayashi, Yoshiharu Toriumi, Yuji Mori, and Hideki Hashimoto. Crystal structure of GeTe and Ge2Sb2Te5 metastable phase. *Thin Solid Films*, 370(1–2):258 – 261, 2000. ISSN 0040-6090.

[185] Volker L. Deringer and Richard Dronskowski. DFT studies of pristine hexagonal Ge1Sb2Te4(0001), Ge2Sb2Te5(0001), and Ge1Sb4Te7(0001) surfaces. *The Journal of Physical Chemistry C*, 117(29):15075–15089, 2013.

[186] H. Krakauer, M. Posternak, and A. J. Freeman. Linearized augmented plane-wave method for the electronic band structure of thin films. *Phys. Rev. B*, 19:1706–1719, Feb 1979.

[187] Bertold Rasche, Anna Isaeva, Alexander Gerisch, Martin Kaiser, Wouter Van den Broek, Christoph T. Koch, Ute Kaiser, and Michael Ruck. Crystal growth and real structure effects of the first weak 3D stacked topological insulator Bi14Rh3I9. *Chemistry of Materials*, 25(11): 2359–2364, 2013.

[188] A L Efros and B I Shklovskii. Coulomb gap and low temperature conductivity of disordered systems. *Journal of Physics C: Solid State Physics*, 8(4):L49, 1975.

[189] T. Mashoff, M. Pratzer, V. Geringer, T. J. Echtermeyer, M. C. Lemme, M. Liebmann, and M. Morgenstern. Bistability and oscillatory motion of natural nanomembranes appearing within monolayer graphene on silicon dioxide. *Nano Letters*, 10(2):461–465, 2010. PMID: 20058873.

[190] Laura Cano-Cortes, Carmine Ortix, and Jeroen van den Brink. Fundamental differences between quantum spin Hall edge states at zigzag and armchair terminations of honeycomb and ruby nets. *Phys. Rev. Lett.*, 111:146801, Oct 2013.

[191] Bin Zhou, Hai-Zhou Lu, Rui-Lin Chu, Shun-Qing Shen, and Qian Niu. Finite size effects on helical edge states in a quantum spin-Hall system. *Phys. Rev. Lett.*, 101:246807, Dec 2008.

[192] Yositake Takane. Conductance of disordered wires with symplectic symmetry: Comparison between odd- and even-channel cases. *Journal of the Physical Society of Japan*, 73(9):2366–2369, 2004.

[193] Chr. Meyer, J. Klijn, M. Morgenstern, and R. Wiesendanger. Direct measurement of the local density of states of a disordered one-dimensional conductor. *Phys. Rev. Lett.*, 91:076803, Aug 2003.

[194] I. Drozdov, A. Alexandradinata, S. Jeon, S. Nadj-Perge, H. Ji, R. J. Cava, B. A. Bernevig, and A. Yazdani. One-dimensional topological edge states of bismuth bilayers. *Nature Phys.*, 10:664–669, 2014.

[195] Fang Yang, Lin Miao, Z. F. Wang, Meng-Yu Yao, Fengfeng Zhu, Y. R. Song, Mei-Xiao Wang, Jin-Peng Xu, Alexei V. Fedorov, Z. Sun, G. B. Zhang, Canhua Liu, Feng Liu, Dong Qian, C. L. Gao, and Jin-Feng Jia. Spatial and energy distribution of topological edge states in single Bi(111) bilayer. *Phys. Rev. Lett.*, 109:016801, Jul 2012.

[196] Michael Ruck. Bi13Pt3I7: Ein Subiodid mit einer pseudosymmetrischen Schichtstruktur. *Zeitschrift für anorganische und allgemeine Chemier anorganische und allgemeine Chemie*, 623(10):1535–1541, 1997. ISSN 1521-3749.

[197] Klaus Koepernik and Helmut Eschrig. Full-potential nonorthogonal local-orbital minimum-basis band-structure scheme. *Phys. Rev. B*, 59: 1743–1757, Jan 1999.

[198] John P. Perdew and Yue Wang. Accurate and simple analytic representation of the electron-gas correlation energy. *Phys. Rev. B*, 45:13244–13249, Jun 1992.

[199] C. Brüne, A. Roth, H. Buhmann, E. M. Hankiewicz, L. W. Molenkamp, J. Maciejko, X. L. Qi, and S-C. Zhang. Spin polarization of the quantum spin hall edge states. *Nature Phys.*, 8:485–490, 2012.

[200] J. Alicea. New directions in the pursuit of Majorana fermions in solid state systems. *Rep. Prog. Phys.*, 75:076501, 2012.

[201] J. Alicea, Y. Oreg, G. Refael, F. von Oppen, and M. P. A. Fisher. Non-abelian statistics and topological quantum information processing in 1D wire networks. *Nature Phys.*, 7:412–417, 2011.

Publications

Publications described within this thesis

1. C. Pauly, G. Bihlmayer, M. Liebmann, M. Grob, A. Georgi, D. Subramaniam, M. R. Scholz, J. Sánchez-Barriga, A. Varykhalov, S. Blügel, O. Rader, and M. Morgenstern. Probing two topological surface bands of Sb_2Te_3 by spin-polarized photoemission spectroscopy. *Phys. Rev. B*, **86**, 235106, 2012.

2. C. Pauly, M. Liebmann, A. Giussani, J. Kellner, S. Just, J. Sánchez-Barriga, E. rienks, O. Rader, R. Calarco, G. Bihlmayer, and M. Morgenstern. Evidence for topological band inversion of the phase change material $Ge_2Sb_2Te_5$. *Appl. Phys. Lett.*, **103**, 243109, 2013.

3. C. Pauly, B. Rasche, K. Koepernik, M. Liebmann, M. Pratzer, M. Richter, J. Kellner, M. Eschbach, B. Kaufmann, L. Plucinski, C. M. Schneider, M. Ruck, J. van den Brink, and M. Morgenstern. Sub-nm wide electron channels protected by topology. *Nature Phys.*, **11**, 338, 2015.

4. C. Pauly, S. Saunus, M. Liebmann, and M. Morgenstern. Spatially resolved Landau level spectroscopy of the topological Dirac cone of bulk-type $Sb_2Te_3(0001)$: Potential fluctuations and quasiparticle lifetime. *Phys. Rev. B*, **92**, 85140, 2015.

Further publications

1. C. Pauly, M. Grob, M. Pezzotta, M. Pratzer, and M. Morgenstern. Gundlach oscillations and Coulomb blockade of Co nanoislands on MgO/Mo(100) investigated by scanning tunneling spectroscopy at 300 K. *Phys. Rev. B*, **81**, 125446, 2010.

2. D. Subramaniam, C. Pauly, M. Liebmann, M. Woda, P. Rausch, P. Merkelbach, M. Wuttig, and M. Morgenstern. Scanning tunneling microscopy and spectroscopy of the phase change alloy $Ge_1Sb_2Te_4$. *Appl. Phys. Lett.*, **81**, 103110, 2009.

3. D. Subramaniam, F. Libisch, Y. Li, C. Pauly, V. Geringer, R. Reiter, T. Mashoff, M. Liebmann, J. Burgdoerfer, C. Busse, T. Michely, R. Mazzarello, M. Pratzer, and M. Morgenstern. Wave-Function Mapping of Graphene Quantum Dots with Soft Confinement. *Phys. Rev. Lett.*, **108**, 046801, 2012.

4. Y. Li, D. Subramaniam, N. Atodiresei, P. Lazic, V. Caciuc, C. Pauly, A. Georgi, C. Busse, M. Liebmann, S. Blügel, M. Pratzer, M. Morgenstern, and R. Mazzarello. Absence of edge states in covalently bonded zigzag edges of graphene on Ir(111). *Adv. Mater.*, **25**, 1967, 2013.

Printed in the United States
By Bookmasters